marscademy.com

(videos, topic-based questions, exam papers and much more)

Published by MarsCademy Publishing

Proofreader: Magdalena Galka

With special thanks to Sunjeong Shin for her support and wisdom.

ISBN: 978-1-915811-03-5

1 About the author

Hello there! I'm Marsida Horeshka, but you can call me Mars, just like the planet – hence the name MarsMaths. With over a decade of experience in the education sector, including several years as a subject leader and manager, I've dedicated my career to making maths more accessible and enjoyable for everyone.

You might already know me from my YouTube channel, Mars Maths, where I post educational videos to help demystify complex maths concepts. I also regularly update my website, marscademy.com, with a variety of resources, including practice questions, answers, and step-by-step solutions to exam papers. I'm constantly adding new content, so be sure to check in regularly for the latest updates.

I'm thrilled to have you join me on this mathematical journey. Together, we'll explore the fascinating world of maths and discover the power and beauty of numbers.

2 About the book

Start by identifying the areas you want to improve. Study the topic, practice the exercises and check your answers. If you need further explanation, you can find my detailed video tutorials on YouTube. Just search for the topic followed by 'Mars Maths'.

Here's what you can expect from this book:

- Step-by-step Explanations
- Visual Representations
- Revisiting Earlier Topics
- Practice Questions
- Practice Paper

This book is designed to help you understand, apply and master maths.

3 You can

It is quite common to feel like you're not doing very well in maths or that you are not a 'maths person'. What I can say is that you're not the only one experiencing these feelings towards maths. If you want to do maths, then you CAN do maths. There is nothing to stop you. Scientific evidence shows that there is no such thing as a 'maths person' and anyone can do maths.

Struggling to complete a maths task though, is a good thing. You are learning if you struggle and if something was too easy, we wouldn't be doing it anyway. Why would we?

Challenge yourself to achieve L2, the level required for many jobs and further/higher education studies. I absolutely believe that you can. Can you? **I challenge you.** And once you've passed your exam, come and tell me on YouTube or social media. I'd love to celebrate your success.

CONTENTS

Self-assessment

Add your confidence level (1 to 5) for each topic before you start studying the topic and when you finish. 1 is the lowest level and 5 is the highest.

Number

- Count using numbers up to 1000
- Read and write numbers up to 1000
- Order and compare numbers up to 1000
- Recognise and continue sequences of whole numbers up to up to 100
- Add three-digit numbers
- Subtract three-digit numbers
- Multiply two-digit by one and two-digit numbers
- Divide three-digit by one and two-digit numbers
- Round to the nearest 10 and 100
- Use thirds, quarters, fifths and tenths and equivalent forms
- Use decimals with up to two decimal places
- Recognise and continue sequences with decimals

Measures, Shape and Space

- Calculate money in pounds and pence
- Round amounts of money to the nearest pound or 10p
- Read, measure and record time using AM and PM
- Read time in hours and minutes
- Compare metric measures of length
- Compare metric measures of weight
- Compare metric measures of capacity
- Use and compare measures to the nearest labelled or unlabelled division
- Use a suitable instrument for measuring length and mass
- Sort 2D and 3Dshapes using their properties
- Describe position and direction

Handling Data

- Extract information from lists
- Extract information from tables
- Extract information from diagrams
- Extract information from bar charts
- Create frequency tables
- Interpret information from bar charts and line graphs
- Represent information in tables, diagrams, bar charts and line graphs

Practice Paper

Count, read, write, order and compare numbers up to 1000

We use numbers when:

- catching a bus
- writing the home address
- reading speed limits
- comparing prices
- buying and selling items
- paying bills

Count using numbers up to 1000

Count in ones up to 1000

These are the numbers from 1 to 100.

1 one	2 two	3 three	4 four	5 five	6 six	7 seven	8 eight	9 nine	10 ten
11 eleven	12 twelve	13 thirteen	14 fourteen	15 fifteen	16 sixteen	17 seventeen	18 eighteen	19 nineteen	20 twenty
21 twenty one	22 twenty two	23 twenty three	24 twenty four	25 twenty five	26 twenty six	27 twenty seven	28 twenty eight	29 twenty nine	30 thirty
31 thirty one	32 thirty two	33 thirty three	34 thirty four	35 thirty five	36 thirty six	37 thirty seven	38 thirty eight	39 thirty nine	40 forty
41 forty one	42 forty two	43 forty three	44 forty four	45 forty five	46 forty six	47 forty seven	48 forty eight	49 forty nine	50 fifty
51 fifty one	52 fifty two	53 fifty three	54 fifty four	55 fifty five	56 fifty six	57 fifty seven	58 fifty eight	59 fifty nine	60 sixty
61 sixty one	62 sixty two	63 sixty three	64 sixty four	65 sixty five	66 sixty six	67 sixty seven	68 sixty eight	69 sixty nine	70 seventy
71 seventy one	72 seventy two	73 seventy three	74 seventy four	75 seventy five	76 seventy six	77 seventy seven	78 seventy eight	79 seventy nine	80 eighty
81 eighty one	82 eighty two	83 eighty three	84 eighty four	85 eighty five	86 eighty six	87 eighty seven	88 eighty eight	89 eighty nine	90 ninety
91 ninety one	92 ninety two	93 ninety three	94 ninety four	95 ninety five	96 ninety six	97 ninety seven	98 ninety eight	99 ninety nine	100 one hundred

After 100, we go to 101, 102, 103, 104, 105, 106, 107, 108, 109, 110, 111 ... 199, 200.

After 200, we go to 201, 202, 203, 204, 205, 206, 207, 208, 209, 210, 211 ... 299, 300.
The same pattern continues after 300, 400, 500, 600, 700, 800 and 900. After 999 comes 1000.

1 Fill this table in with numbers from 101 to 200.

101									110
191									200

2 Fill the gaps.

301									310
									320
									330
									340
									350
									360
									370
									380
									390
391									400

3 Fill the gaps.

703		705		707		709		711	
450		452		454		456		458	
	643		645		647		649		651
	886		888				892		894
225		227		229		231			
	550		552				556		
888	889				893				
		167		169					
561									570
307									316
792					797				

Counting in twos up to 1000

This is how we count in twos.

+2 +2 +2 +2

34 → 36 → 38 → 40 → 42

4 Fill the gaps.

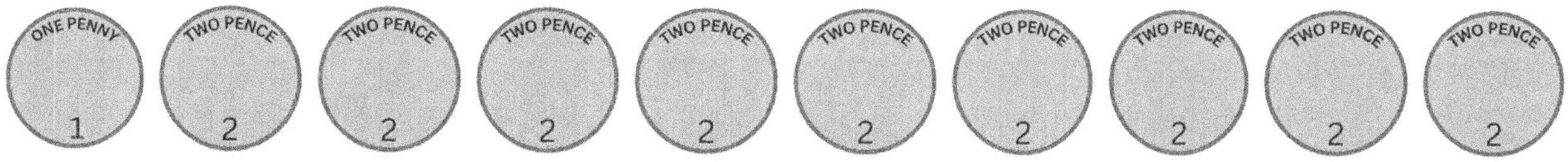

1p 3p 5p ___p 9p ___p ___p ___p ___p ___p

5 Complete this number pattern by counting in twos.

		84		88		92		96	
	forty nine			fifty five					
	163	165		169		173		177	
		two hundred and eighty five				two hundred and ninety three		two hundred and ninety seven	
572		576		580				588	
	eight hundred and fifty four		eight hundred and fifty eight			eight hundred and sixty four			

6 Fill the gaps by counting in twos.

112, 114, ___ 118, ___, 122, ___	751, 753, ___, 757, ___, ___, ___	520, ___, 524, ___, 528, ___, 532
230, ___, 234, 236, ___, ___	565, ___, ___, ___, 573, ___	688, ___, ___, 694, ___, ___
___, ___, 444, 446, ___, ___	372, ___, ___, 378, ___	890, ___, ___, ___, ___, 900

7 Fill the gaps, counting in ones, twos and backwards.

330, 331, ___, 333, ___, 335, ___	___, ___, 684, 682, ___, 678	542, ___, 546, ___, ___, ___
750, ___, 754, ___, ___, 760	275, ___, ___, 281, 283, ___	468, ___, ___, 474, ___, ___

Odd and even numbers

Door numbers can be odd or even.

Odd numbers go like this: 1, 3, 5, 7, 9, 11, 13, 15...

Even numbers go like this: 2, 4, 6, 8, 10, 12, 14...

Houses on one side of the road have odd numbers. For example: 13, 15, 17, 19.
Houses on the other side of the road have even numbers. For example: 14, 16, 18, 20.

Add the missing numbers.

165 ___ 169 ___

8 Are these numbers even or odd?

73 odd	9 ___	92 ___	35 ___	14 ___	516 ___
452 ___	45 ___	52 ___	98 ___	37 ___	20 ___
648 ___	857 ___	210 ___	649 ___	698 ___	243 ___

To decide if 156 is odd or even, you only need to look at the ones: 6. 6 is even, so the whole number is even. 593 is odd because 3 is odd.

9 a) Write an even number that is greater than 200. ___

b) Write an odd number that is smaller than 600. ___

Counting in 10s up to 1000

This is how we count in tens.

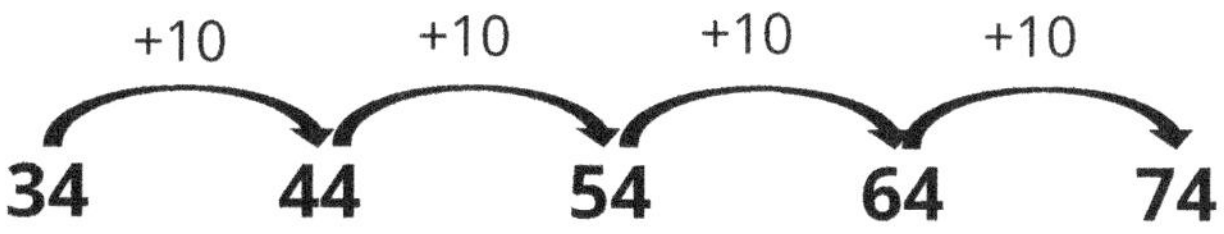

10 Fill the gaps.

10 ___ 30 ___ 50 ___ ___

11 Complete this number pattern by counting in tens.

		80		100		120		140	
	forty nine			seventy nine					
	163	173		193		213		233	
		two hundred and eighty five				three hundred and twenty five		three hundred and forty five	
572		592		612				652	
	Eight hundred and fifty four		Eight hundred and seventy four			Nine hundred and four			

12 Noor works in a shop. She checked in the morning and there was £245 available in the shop. A customer bought an item and paid with these notes.

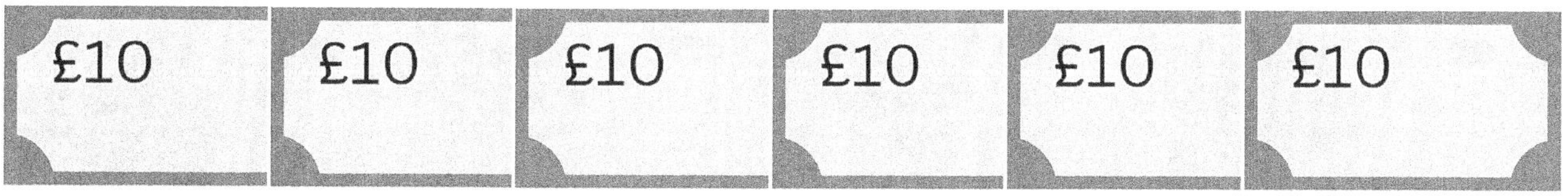

How much money is now available in the shop? ______________________________

13 Bo was given some money in £10 notes for his birthday. Before the birthday he had £152. After the birthday he has £232. How many £10 notes was he given? ______________________________

Counting in 100s up to 1000

This is how we count in 100s.

14 Fill the gaps.

100p 200p 300p _____p 500p _____p _____p _____p _____p

15 Complete this number pattern by counting in hundreds.

80		280		480		680		880	

	one hundred and forty nine			four hundred and forty nine					

	163	263		463		663		863	

		two hundred and eighty five				six hundred and eighty five		eight hundred and eighty five	

16 Rupinder deposits some €100 notes into her bank account. Her new account balance is €937. Her initial balance was €237.

How many €100 notes did Rupinder deposit into her account?

____

17 Sal is working out how much money these notes make.

How much are these notes worth? €______

18 Janey is the manager of a grocery store. She is checking stock levels for fruit and vegetables.

a) There are already 45kg tomatoes in store. These crates have arrived. Each crate holds 10kg of tomatoes.

How many kg of tomatoes are available in store after the delivery? ______________________

b) There are already 113 kg pears in store. These crates have arrived. Each crate holds 20kg of pears.

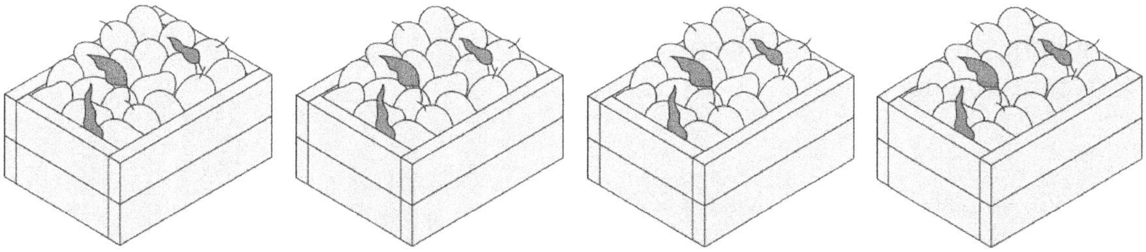

How many kg of pears are available in the store after delivery? ______________________

19 A charity distributes boxes with essentials to people in need. In each box there are 100 items. The charity has already given away 450 items.

How many items has the charity given away, including these three boxes?

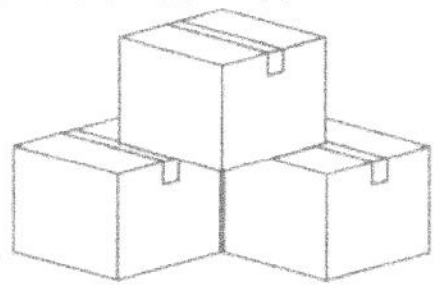

____

20 Siobhan says 'Counting in 10s is easy as numbers end in 0'.

a) Is Siobhan correct? ______________________

b) Is this always the case? ______________________

21 Sal says 'Counting in 100s is also easy as numbers end in 0'.

a) Is Sal correct? ______________________

b) Is this always the case? ______________________

22 The house numbers on one side of the road go like this: **235, 237, 239, 241, 243 ...**

Will 264 be on the same side of the road? ______________________

23 A plane has seats numbered from 1 to 240. The even-numbered seats are by the window. Sam is sitting in seat number 138. Is she sitting by the window? Why?

____

24 Fill the gaps.

7 + 5 = ___	15 + 9 = ___	23 + 17 = ___	➡	odd	+	odd	=	even
6 + 8 = ___	10 + 14 = ___	26 + 26 = ___	➡	even	+	even	=	_____
9 + 12 = ___	11 + 6 = ___	21 + 34 = ___	➡	odd	+	even	=	_____
6 + 7 = ___	12 + 7 = ___	36 + 13 = ___	➡	even	+	odd	=	_____

Progress check

1 Ann is counting to 1000. She missed some numbers. Add the missing numbers in the sequence.

245, _____, 247, 248, _____, _____

2 Bieta is counting the value of her £2 coins. She goes: _____, 112, 114, 116, 118, _____
Add the two missing numbers.

3 Sidra has £350 in cash. She withdraws six £50 notes.

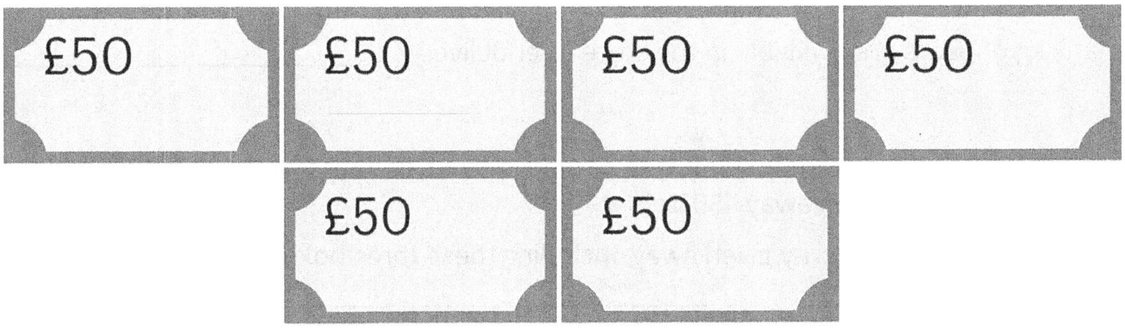

How much money does Sidra have now? ______________________________

4 Fill in the blank spaces in this number pattern.

_____, _____, 710, 720, 730, 740, 750

5 Put these numbers in order, starting with the largest.

52 **103** **250** **130** **0** **100**

_____, _____, _____, _____, _____, _____

6 Jane completes a number pattern, like below. Is she right? Why?

180, 190, 200, 201

____

Read and write numbers up to 1000

Representing numbers up to 1000

This represents 110.

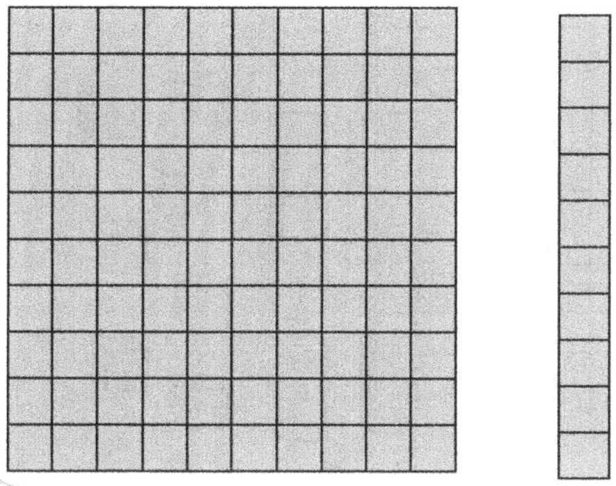

This represents 130.

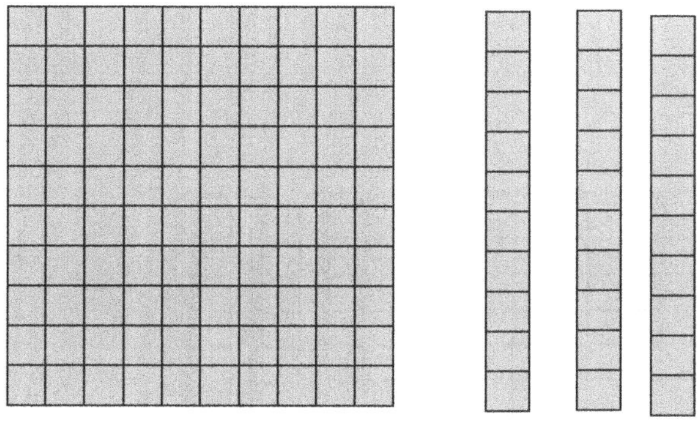

This represents 138.

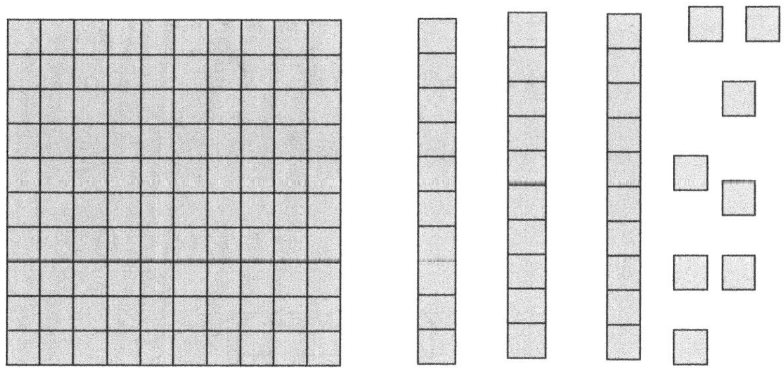

4 is a number.

This number has **one digit.**

4 means **4**

69 is a **number**.

This number has **two digits**:

6 means **60**

9 means **9**

69 = 60 + 9

197 is a **number**.

This number has **three digits**:

1 means **100**

9 means **90**

7 means **7**

197 = 100 + 90 + 7

642 is a **number**.

This number has **three digits**:

6 means **600**

4 means **40**

2 means **2**

642 = 600 + 40 + 2

This represents 244.

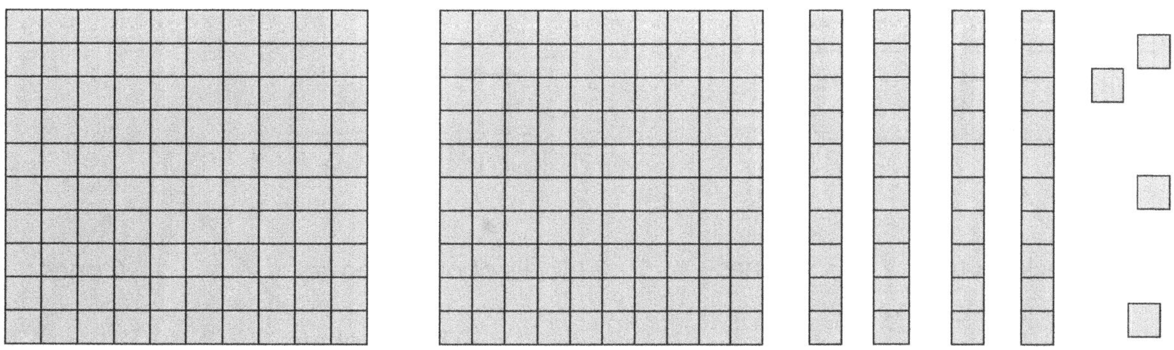

Hundreds, tens and ones

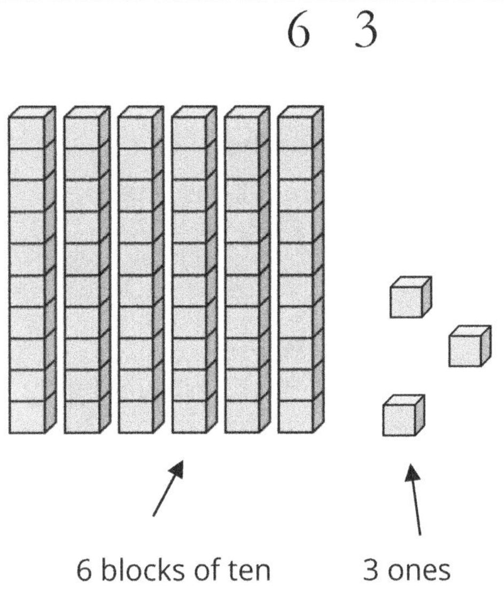

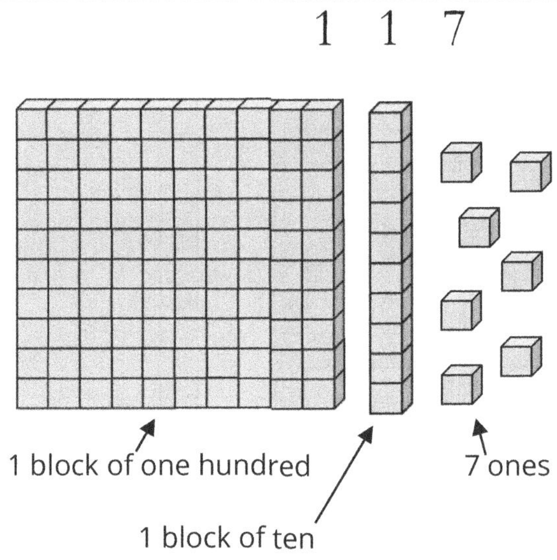

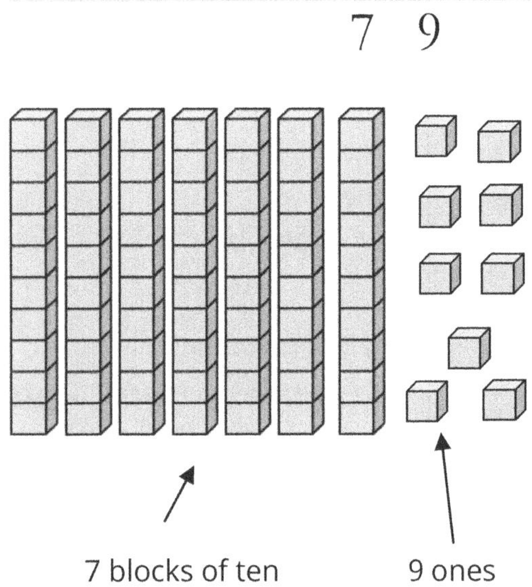

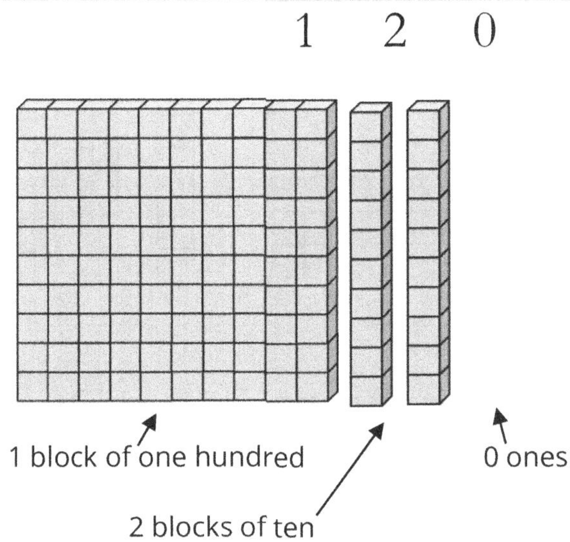

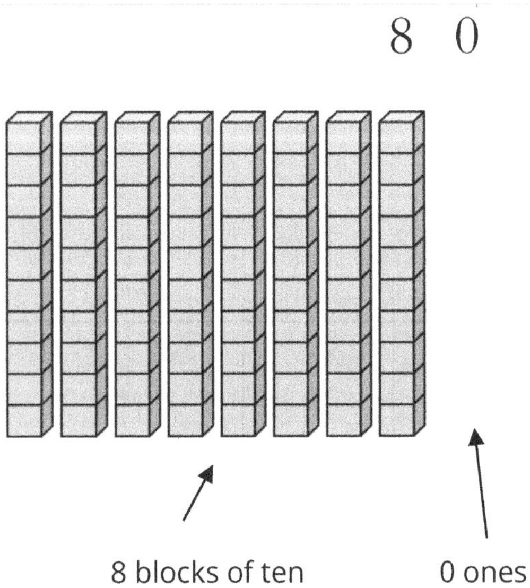

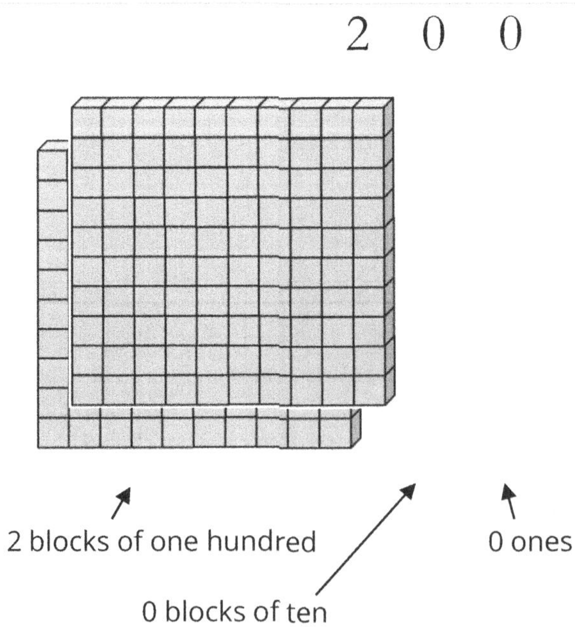

Key Words

Another name for **ones** is **units.**
Number 54 has **5 blocks of 10** and **4 units.** Number 78 has **7 blocks of 10** and **8 units.**

1 Numbers 68 and 150 have been added to the grid below.

Hundreds	Tens	Units
	6	8
1	5	0

Add these numbers to the grid: 39, 470, 402, 758, 300, 407, 836, 999.

Read and write numbers up to 1000

2 Write these numbers in **digits**.

one hundred and eleven ______	ninety nine ______
one hundred and twenty ______	seven hundred and seventy one ______
one hundred and sixty ______	three hundred and eighteen ______
two hundred and seventy five ______	one hundred and twenty four ______
four hundred and thirty four ______	five hundred and seven ______
nine hundred and eighty six ______	two hundred ______
eight hundred and eighty six ______	six hundred and eighty six ______

3 Write the number that has eight hundreds, three tens and five ones. ______

4 Add the missing numbers. The first one has been done for you.

The Bank App

Amount you are sending in digits:

£ 179 ______________

in words:

one hundred and

seventy nine

_____pounds

The Bank App

Amount you are sending in digits:

£ ______________

in words:

five hundred and

thirty eight

_____pounds

The Bank App

Amount you are sending in digits:

£ ______________

in words:

four hundred

____

_____pounds

The Bank App

Amount you are sending in digits:

£ 844 ______________

in words:

____

____

_____pounds

The Bank App

Amount you are sending in digits:

£ 609 ______________

in words:

____

____

_____pounds

The Bank App

Amount you are sending in digits:

£ 970 ______________

in words:

____

____

_____pounds

5 Sally has invited her friend over. She explains on the phone: 'I live at four hundred and two White Street.' Write Sally's address using digits.

____

6 The population of a town is **eight hundred and fifty nine**. Write the population of the town in digits.

____

7 Write the number that has two hundreds, zero tens and nine ones. ______________

8 Write these numbers in **words**.

67 ______

113 ______

120 ______

646 ______

258 ______

819 ______

790 ______

366 ______

184 ______

500 ______

407 ______

938 ______

9 Three people are waiting in the queue in a shop. Fill the gap by writing the number **in words**.

Heather Jasi Enes

Jasi is number ______ in the queue.

10 Hamza has completed this deposit form, paying **eight hundred and twenty** pounds.

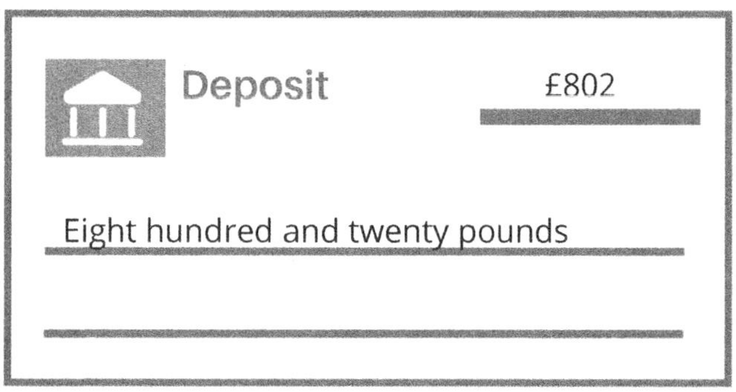

What has he done wrong? ______

11 Lou received her gas bill in the post.

a) How much does she have to pay in total? Write your answer in words.

____

b) Lou's bank balance is £379. Does she have enough to pay the bill?

____

Gas Bill

January £119
February £107
March £100
TOTAL £326

12 Les is taking out a loan. He will pay back **six hundred ninety-eight** pounds every week.
Write this amount in numerals.

____

Another two words that mean the same as **digits** are **figures** and **numerals.**

Progress check

1 Numbers 68 and 150 have been added to the grid below.

Hundreds	Tens	Units
	6	8
1	5	0

Add these numbers to the grid: 141, 208, 290.

2 Write these numbers in words.

Number in digits	Number in words
117	
890	
601	
1000	

3 Ayla writes the number 118 as 'one hundred and eighty'. Is she right? Explain your answer.

____

4 How do you write the number 555 in words?

____

5 Jane fills in the cheque below. Write the amount in digits in the cheque.

The Bank

Pay Joe Bloggs

five hundred and twenty pounds only

£

Order and compare numbers up to 1000

Compare numbers up to 1000

Here are the numbers from 0 to 1000 on a number line. We can use a number line to compare numbers.

300 is **greater** than 200

550 is **less** than 600

1 Fill these gaps with **greater** or **less**.

442 is ________ than 400

674 is ________ than 670

370 is ________ than 307

780 is ________ than 870

2 Fill these gaps with **greater** or **less**.

319 is ________ than 300	780 is ________ than 708	560 is ________ than 650
560 is ________ than 530	459 is ________ than 549	1000 is ________ than 0
628 is ________ than 682	1000 is ________ than 999	640 is ________ than 365
417 is ________ than 407	779 is ________ than 799	349 is ________ than 358

When comparing **numbers that are greater than 100 with numbers that are less than 100**, the number that is greater than 100 is greater than the other one: **183 is greater than 83**

If **both numbers are greater than 100**, the number with more tens is greater: **183 is greater than 173**

If **both numbers are greater than 100** and **they have the same number of tens**, the number with more ones is greater: **183 is greater than 182**

3 Complete these sentences using **higher** or **lower**.

429 is ________ than 329	800 is ________ than 776	886 is ________ than 894
515 is ________ than 550	654 is ________ than 456	335 is ________ than 366
108 is ________ than 180	659 is ________ than 655	95 is ________ than 876
670 is ________ than 760	769 is ________ than 770	755 is ________ than 366

4 a) Find a number that is greater than 276. ______________________

b) Find a number that is smaller than 460. ______________________

c) Find a number that is greater than 276 but smaller than 460. ______________________

Greatest and **smallest**

Out of **500**, **550** and **600**, 500 is **the smallest** as it is the first in the number line.

Out of **510**, **260** and **380**, 510 is **the greatest** as it is the last in the number line.

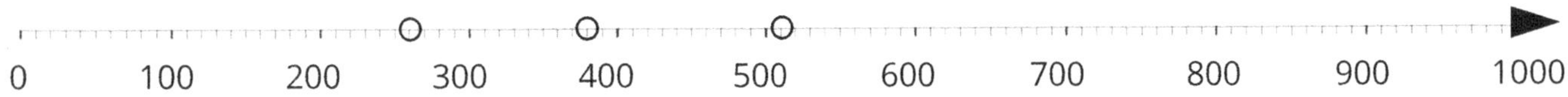

5 Fill the gaps.

Out of **830**, **880** and **900**, _____ is **the smallest** and _____ **is the greatest**.

Out of **670**, **530** and **1000**, _____ is **the smallest** and _____ **is the greatest**.

Out of **400**, **150** and **999**, _____ is **the smallest** and _____ **is the greatest**.

6 a) Circle the number that is the **greatest**:
784, 780, 789, 707

b) Circle the number that is the **greatest**:
220, 300, 420, 195

c) Circle the number that is the **greatest**:
830, 803, 83, 380

d) Circle the number that is the **greatest**:
475, 298, 478, 160

e) Circle the number that is the **greatest**:
276, 937, 265, 167

f) Circle the number that is the **smallest**:
74, 7, 107, 710

g) Circle the number that is the **smallest**:
220, 300, 420, 189

h) Circle the number that is the **smallest**:
830, 803, 83, 380

i) Circle the number that is the **smallest**:
475, 298, 478, 645

j) Circle the number that is the **smallest**:
276, 937, 265, 804

Order numbers up to 1000

7 Put these numbers in order, **starting with the smallest**.

392, 367, 391 ______, ______, ______	400, 40, 140 ______, ______, ______
274, 280, 295 ______, ______, ______	818, 180, 108 ______, ______, ______
672, 600, 620 ______, ______, ______	169, 145, 137 ______, ______, ______

8 Put these numbers in order, **starting with the greatest.**

684, 657, 603 ______, ______, ______	174, 187, 153 ______, ______, ______
163, 496, 200 ______, ______, ______	384, 386, 353 ______, ______, ______
879, 840, 804 ______, ______, ______	969,970, 971 ______, ______, ______

9 Put these measures in order starting with the highest: 478m 170m 380m 187m

________, ________, ________, ________

10 Sonia is looking for a cheap flight to Australia. These are the prices of the tickets she found online.

Ticket A £835	**Ticket B** £853	**Ticket C** £805	**Ticket D** £899

a) Which airline sells the cheapest ticket? ______________________________

b) Put the tickets in order, starting with the cheapest: ________, ________, ________, ________

11 Fey is ordering some stationery for the school she works in. She wants to buy the box with the highest number of paper sheets in it. Tick the box Fey will buy.

Box A	**Box B**	**Box C**	**Box D**
500 sheets	485 sheets	750 sheets	650 sheets

12 Write a number that is greater than 790 but smaller than 850. ______________________________

13 Asma wants to buy a washing machine. She will buy the cheapest one.

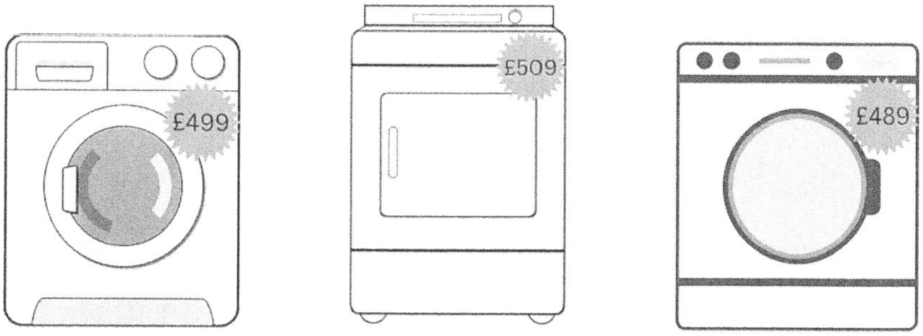

a) How much does the cheapest washing machine cost? ______________________________

b) Put these prices in order, starting with the cheapest: £_____, £_____, £_____

Progress check

1 Put these amounts in order, starting with the one with the highest value: £908 £980 £189 £981

________, ________, ________, ________

2 a) Write the amounts in digits on the cheques.

The Bank

Pay Joe Bloggs

seven hundred and thirty two pounds £

The Bank

Pay Joe Bloggs

seven hundred and thirty pounds £

b) Which of these cheques has the highest value? Tick it.

3 Susan ordered some numbers from the lowest to the highest: 518, 591, 572, 603, 705.
Is she correct? Why?

____

4 Amla is buying a fridge-freezer. She has checked the prices in different shops for the fridge-freezer she wants to buy.

Ample Appliances £429
The Fridges £399
The White Appliances £449

Amla wants to buy the cheapest of them all. Where should she buy the fridge-freezer from?

____

5 Aaron says: 'Four hundred and eighty is bigger than four hundred and eighteen'. Is Aaron correct?
Explain your answer.

____

Recognise and continue linear sequences of numbers up to 100

We come across sequences in:

- clothes sizes
- door numbers
- times tables
- shoe sizes

When counting in ones, twos, threes, fours, fives, sixes or more, you are creating linear sequences. In a linear sequence the numbers go up (e.g. 2, 4, 6, 8) or down (e.g. 12, 9, 6, 3) by the same amount each time.

Clothes sizes go up in a sequence. Example: 10, 12, 14, 16, 18, 20...
Door numbers go up in a sequence. Example: 7, 9, 11, 13, 15...

In this number line, numbers are going up by 4, so 4 is added each time:

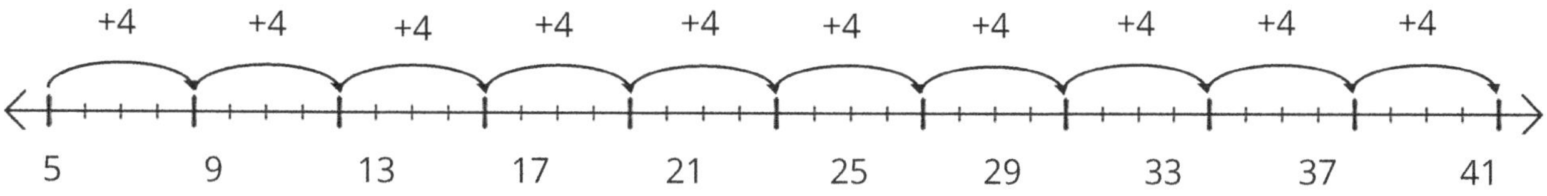

In this number line, numbers are going down by 7, so 7 is subtracted each time:

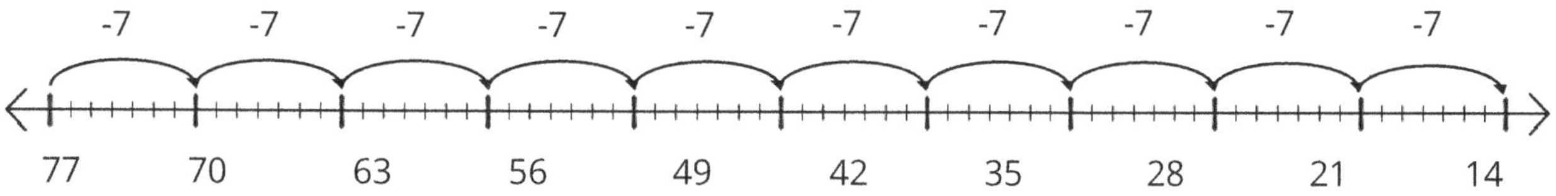

To work out what the sequence is going up or down by, you find the difference between two numbers next to each other.
For this sequence: 3, 8, 13, 18, 23 we do: 8 – 3 = 5 or 13 – 8 = 5 so the sequence increases by 5
For this sequence: 14, 11, 8, 5, 2 we do: 14 – 11 = 3 or 8 – 5 = 3 so the sequence decreases by 3

1 Anna is a teacher. She is buying highlighters for the whole class. In one pack there are 6 highlighters. She counts: 6, 12, 18, ______, ______
Add the next two numbers to the sequence.

2 Lina visits her mother every three days. It is 4th July and she's just visited her mother. She writes down the dates when she will see her mother again.
4th, 7th, 10th, ______, ______, ______
a) Add the next three dates to the sequence.
b) Will she visit her mother on the 20th? Show your working.

____

July						**2023**
Mon	Tue	Wed	Thu	Fri	Sat	Sun
1	2	3	4	5	6	7
8	9	10	11	12	13	14
15	16	17	18	19	20	21
22	23	24	25	26	27	28
29	30					

3 Nesa is buying a water bottle from the vending machine. She has 10p coins only.
She counts as she puts the money into the machine: 10p, 20p, 30p, 40p, _____, _____
a) Add the next two amounts to the sequence
b) The water bottle costs £1. How many 10p coins does she need in total?

____

4 Nita is counting her change. She has 90p in her purse. As she takes a coin each time, she works out how much is left: 90p, 70p, ____, ____, ____
Continue the sequence.

5 Karen works in a bakery. She knows she has 50 strawberries in the fridge. Each time she takes out the same number of strawberries to decorate cakes. She keeps track of how many strawberries are left in the fridge.
50, 46, 42, ______, ______

a) Continue the sequence.

b) How many strawberries is Karen using to decorate each cake?

____

c) How many strawberries will be left over after decorating the maximum number of cakes?

____

6 Times tables are sequences too. Fill the gaps.

2	4	6									
5	10	15									
				35	42	49					
							32	36	40		
		27	36	45							
									60	66	72
8	16	24									
			12	15	18						
						77	88	98	110	121	132

7 Continue these sequences.

2, 4, 6, 8, ___, ___	22, 32, 42, ___, ___	52, 57, 62, ___, ___	0.5, 1, 1.5, ___, ___
37, 35, 33, ___, ___	66, 55, 44, ___, ___	50, 56, 62, ___, ___	3.7, 4.8, 5.9, ___, ___
74, 77, 80, ___, ___	64, 72, 80, ___, ___	8, 16, 24, ___, ___	8.0, 9.2, 10.4, ___, ___

8 Fill the gaps in these sequences.

5, 6, 7, 8, ___, ___, ___	14, 21, 28, ___, ___, 49	72, ___, 76, ___, 80, ___
___, ___, 30, ___, ___, 33	6.6, 5.5, 4.4, ___, ___, 1.1	65, 62, 59, ___, ___, 50
4.5, 4.7, 4.9, ___, ___, 5.5	34, 38, 42, ___, 50, ___, ___	2.9, 2.6, 2.3, ___, ___

9 A sequence goes like this: 28, 25, 22, ... What comes before 28? ______________________

10 Sue works as a shop assistant. She is giving change to a customer.
Counting the coins, she goes: 10p, 20p, 30p, 40p, ________, ________. Fill the gaps.

11 A sequence goes like this: 49, 58, 67, ... Will 94 be in this sequence? Show your working.

____

12 What is the next number in this sequence? 23, 42, 61, 80, _______

13 A sequence goes like this: 2, 4, 8, 16, 32... Is this a linear sequence? Why?

____

14 Continue this sequence: 6.5, 7, 7.5, _______, _______, _______

15 A sequence goes like this: 49, 58, 67, _______ Bardha says: 'This is as easy as it is going up by one within the tens and down by one within the units'.

Do you agree? Explain. ____

Progress check

1 Look at this sequence of numbers: 6, 12, 18, 24, 32

What happens from one number to the next? ______________________________

2 Continue this sequence by adding two numbers in the gaps: 42, 36, 30, 24, _____, _____

3 Fill the gaps in this number pattern: 55, 60, 65, _____, 75, _____, _____, 90

4 Fill in the blank spaces in the pattern below. _____, _____, 24, 28, 32, 36

5 A number pattern starts like this: 9, 18, 27... and goes up to 90. Will number 81 be in this pattern? Why? ____

6 Suki writes this number pattern: 33, 36, 49, 52 Is she correct? Why?

____

7 Guillermo says 'In a sequence, you always need to add or take away the same amount".
Is he correct? Explain.

____

Add and subtract using three-digit numbers

In everyday life, we add and subtract, to work out:

- the cost of items we are buying and the change we will get
- the number of hours spent studying in a week
- the cost of bills or subscriptions per year
- the distance between two places
- the scores in a game
- the cost of organising an event
- the amount of paint you need to paint the walls
- the amount for each ingredient when cooking for many people
- the savings made over a year

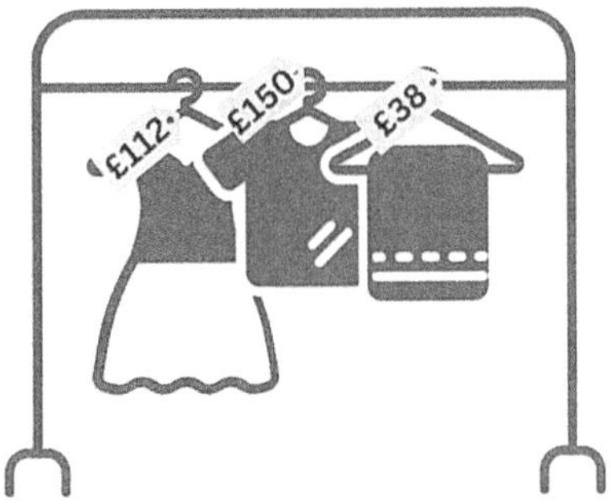

Add three-digit numbers

The part-whole method of addition

This is another method you can use to add.

35 + 78

We split 35 into 30 and 5.

We split 78 into 70 and 8.

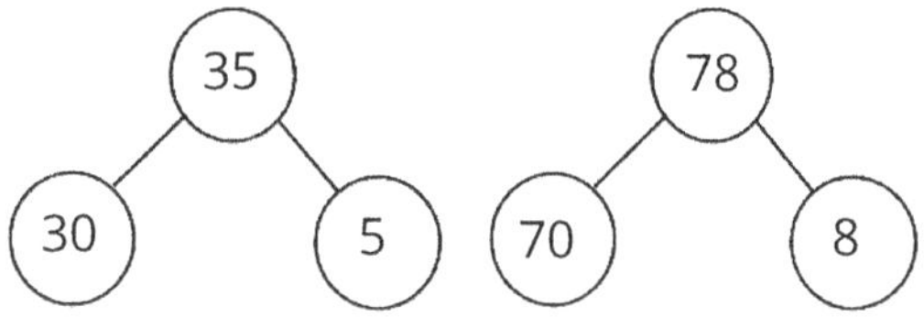

Now we add the tens: 30 + 70 = 100

And the ones: 5 + 8 = 13

Altogether: 100 + 13 = 113

124 + 253

We split 124 into 100 and 20 and 4.

We split 253 into 200 and 50 and 3.

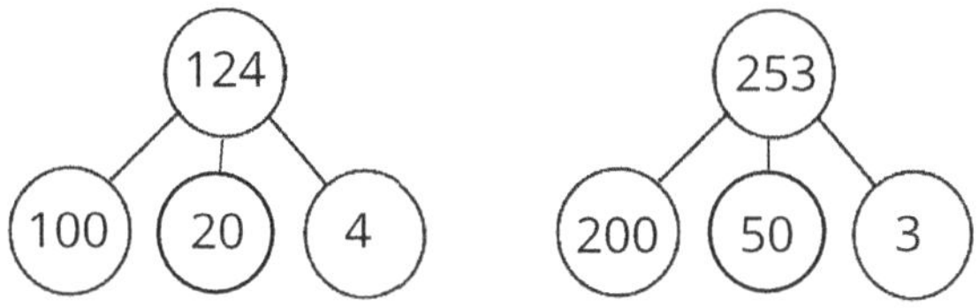

Now we add the hundreds: 100 + 200 = 300

Add the tens: 20 + 50 = 70

And the ones: 4 + 3 = 7

Altogether: 300 + 70 + 7 = 377

1 Fill the gaps.

69 + 43

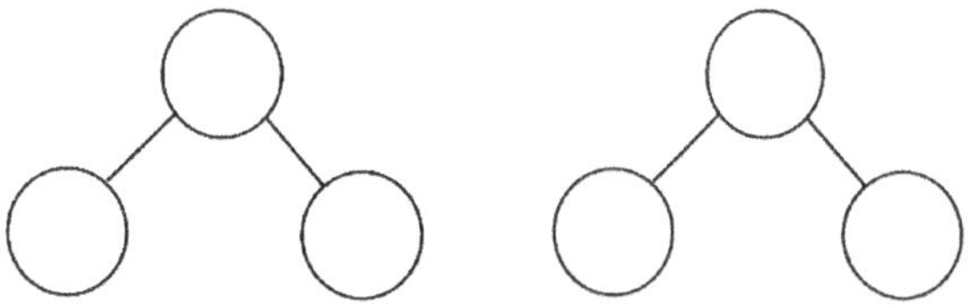

75 + 82

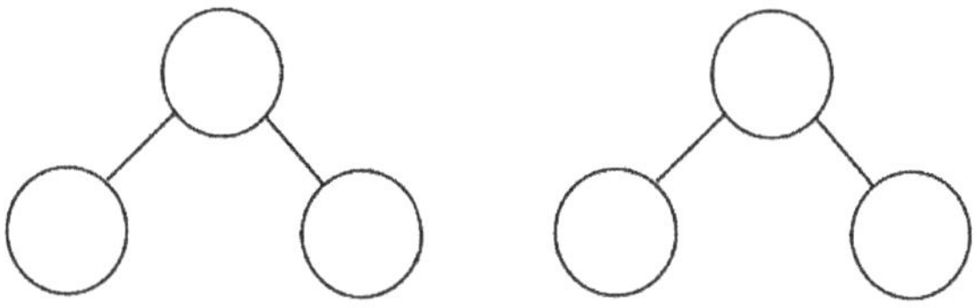

56 + 37

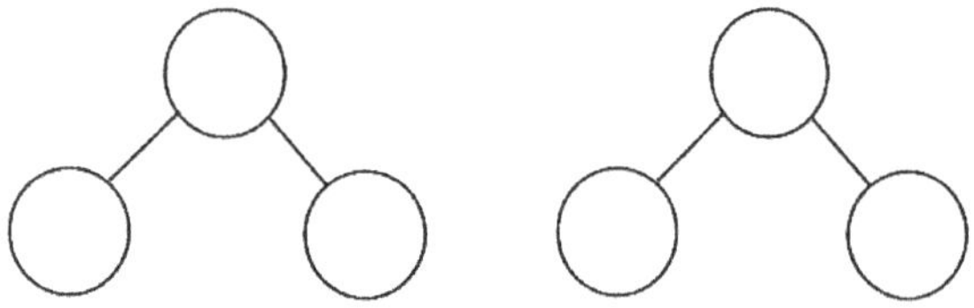

92 + 29

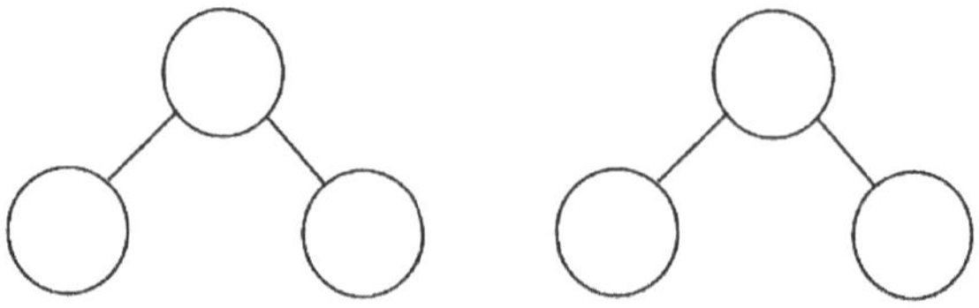

2 Fill the gaps.

6210+ 206

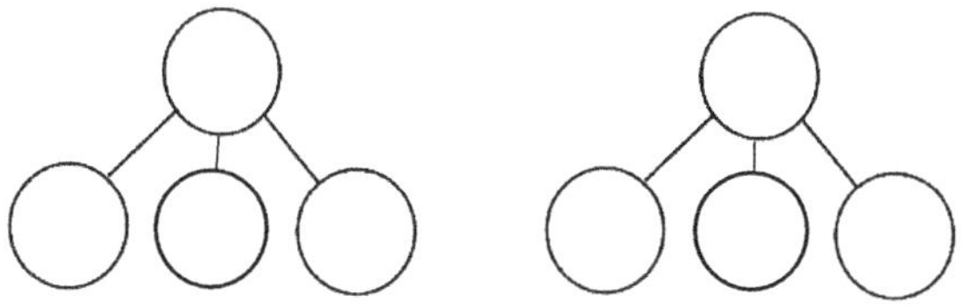

353 + 204

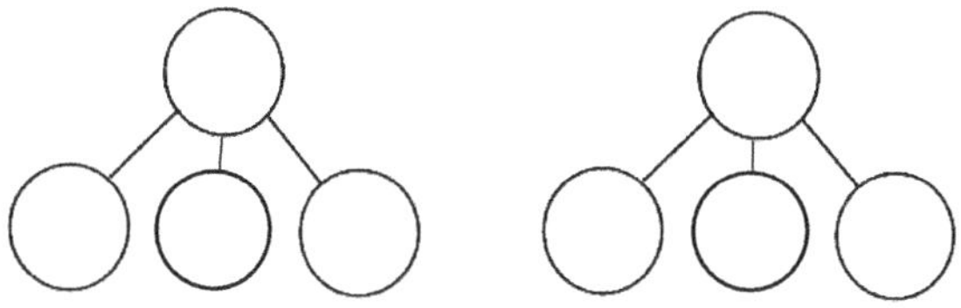

463 + 235

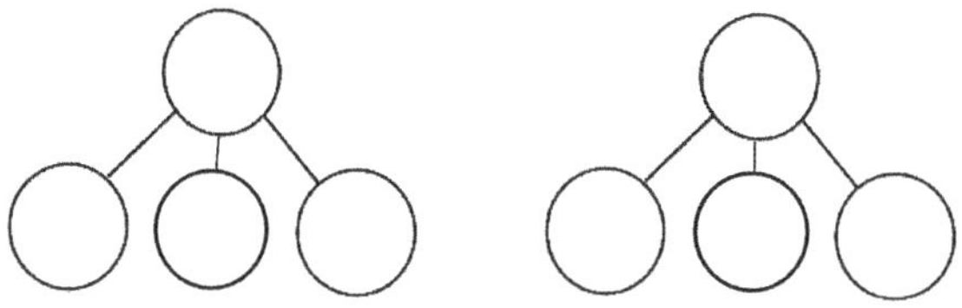

469 + 226

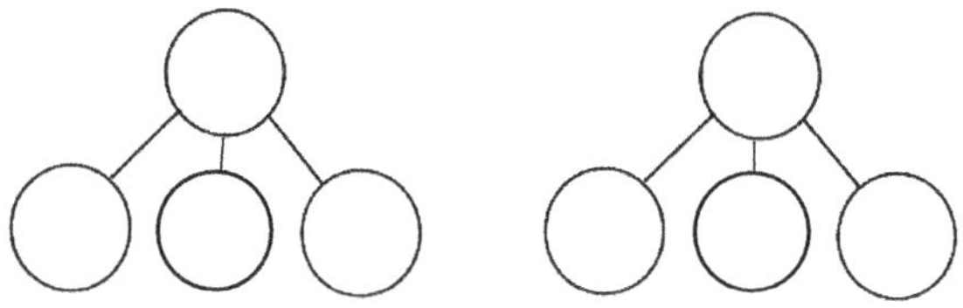

305 + 236

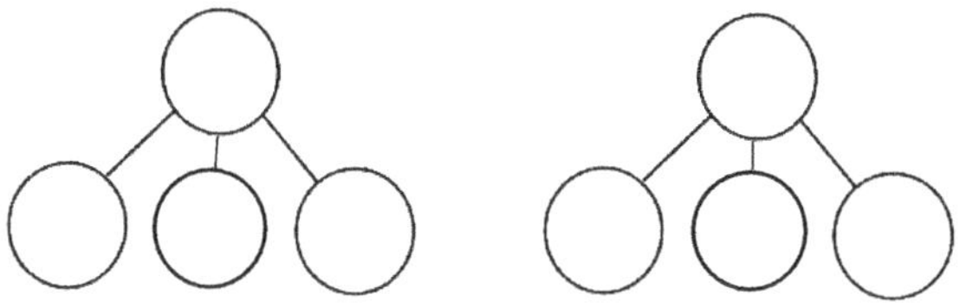

392 + 243

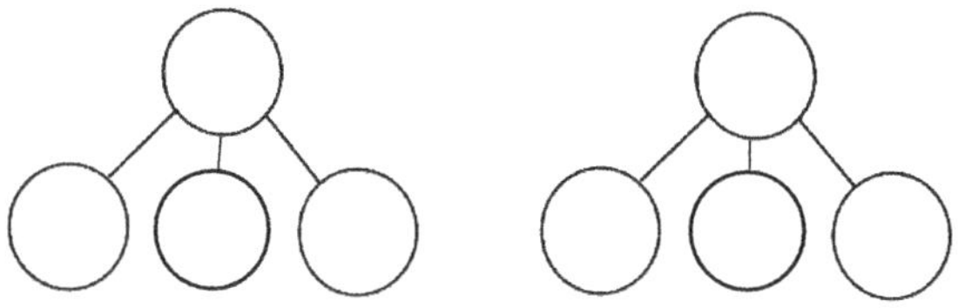

534 + 547

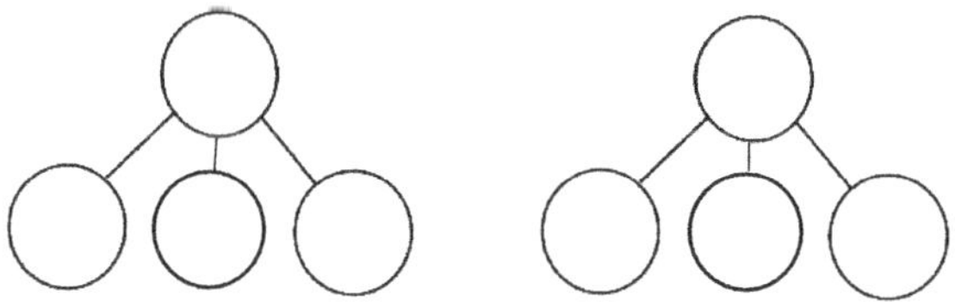

754 + 642

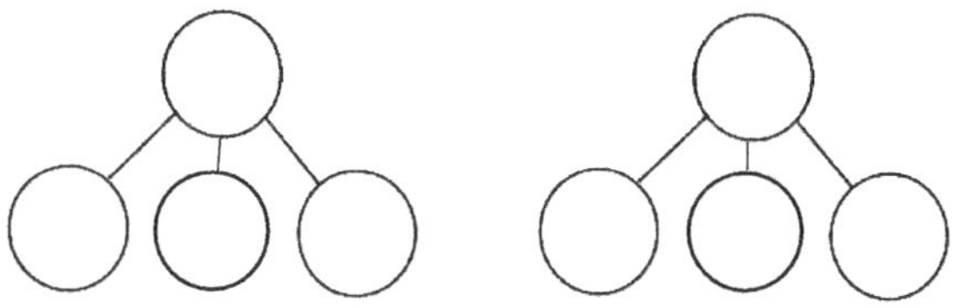

302 + 976

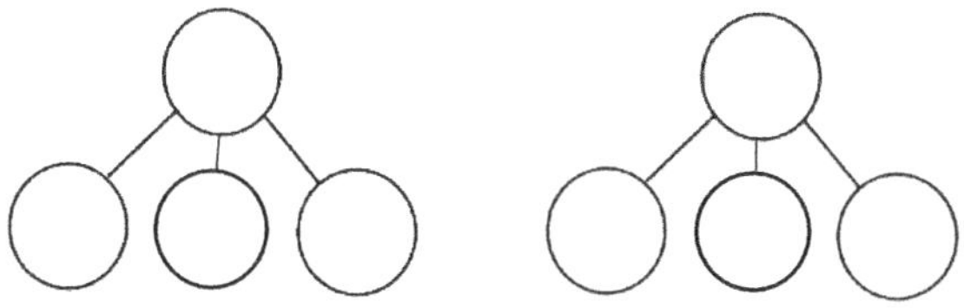

651 + 543

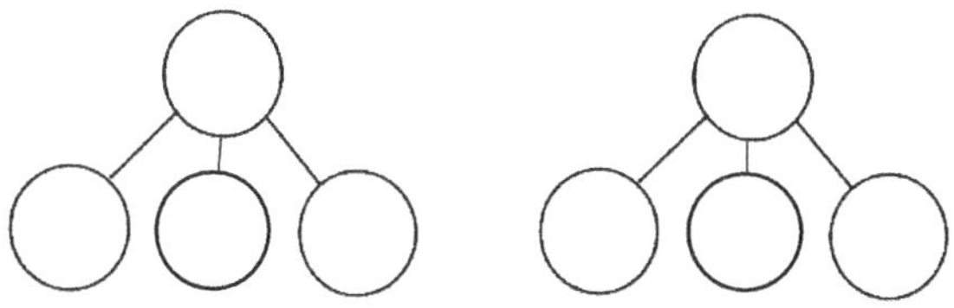

The column method of addition

Example 1

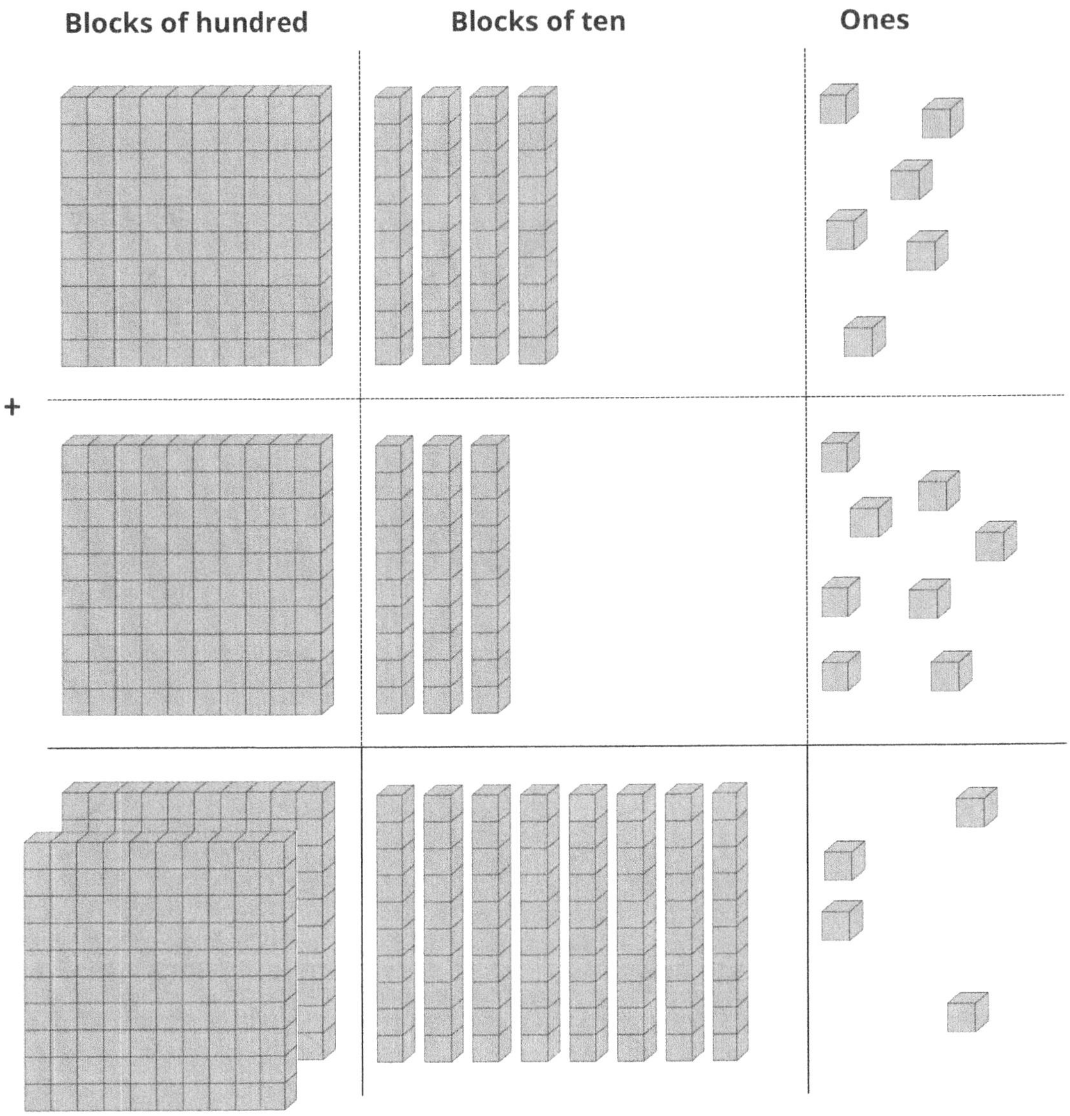

	Hundreds	Ten	Ones
		1 (carried)	
	1	4	6
+	1	3	8
	2	8	4

As 6 + 8 = 14, we carry 10 over to the tens column and put 4 in the ones column.

Example 2

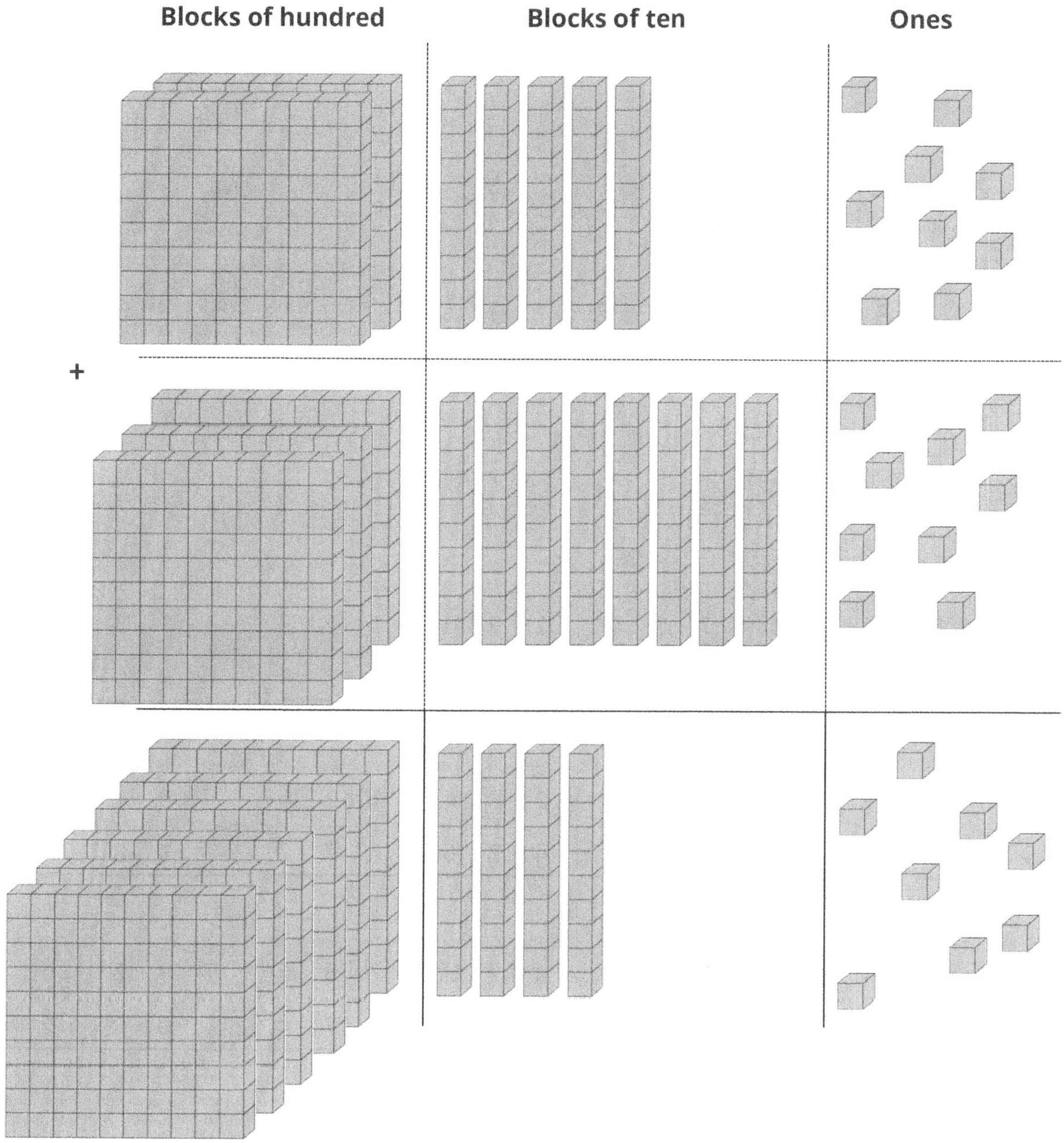

	Hundreds	Tens	Ones
	1	1	
	2	5	9
+	3	8	9
	6	4	8

As 9 + 9 = 18, we carry 10 over to the tens column and put 8 in the ones column.

As 5 + 8 + 1 = 14, we carry 10 over to the hundreds column and put 4 in the tens column.

You can use whichever method you like to add. It is useful to know them all.

Here are some more examples.

		1	
	3	8	5
+	2	4	7
	6	3	2

	1	1	
	7	0	2
+		9	9
	8	0	1

	1	1	
		8	8
+	4	7	5
	5	6	3

3 Fill the gaps.

1 9 1 + 6 7 ___	1 2 4 + 4 8 5 ___	7 5 6 + 8 4 ___	1 9 0 + 6 8 7 ___	1 9 3 + 7 7 3 ___	9 1 3 + 8 6 ___	8 1 4 + 1 5 7 ___
4 0 8 + 5 8 1 ___	5 7 6 + 3 2 7 ___	4 1 5 + 3 7 2 ___	2 8 4 + 4 2 6 ___	1 2 6 + 7 7 ___	9 1 + 7 3 8 ___	6 6 4 + 7 4 ___
6 4 1 + 3 4 8 ___	4 7 0 + 1 5 8 ___	5 5 4 + 1 0 2 ___	4 7 2 + 5 2 7 ___	7 6 5 + 1 7 4 ___	7 9 2 + 1 0 9 ___	5 5 3 + 2 8 3 ___
6 9 5 + 2 0 5 ___	2 3 1 + 1 7 5 ___	3 6 2 + 2 7 6 ___	5 5 9 + 3 1 0 ___	1 1 4 + 3 2 7 ___	1 8 6 + 3 5 7 ___	4 5 8 + 3 9 6 ___
6 4 2 + 2 8 5 ___	1 7 3 + 2 3 5 ___	4 1 6 + 5 7 3 ___	3 7 8 + ___ 9 8 9	4 6 8 + ___ 7 3 2	5 7 4 + ___ 8 7 6	5 8 0 + ___ 7 6 5

Add these numbers.

14 + 145 =	236 + 78 =	616 + 247 =	509 + 174 =
145 + 14 =	78 + 236 =	247 + 616 =	174 + 509 =

What do you notice?

> Order doesn't matter when adding.
> 4 + 3 = 7   3 + 4 = 7

4 Sylvia read 216 pages of a book in the morning and 88 pages in the afternoon. How many pages has she read in total?

___

5 Nadia spent £656 on rent and £239 on food last month. How much did she spend in total?

___

6 Dan's phone has 43GB free storage space. He also has a memory card on the phone with 256GB capacity. How much free storage does Dan's phone have, including the memory card?

___

7 Atif is paying his gas, electricity and water bills.

Gas £149
Electricity £214
Water £96

Atif thinks he will pay less than £500 in total. Is he correct?

___

8 Katheryn says: 'When adding 0 to a number, the number stays the same'. Is Katheryn correct?

___

9 Sarah goes shopping and buys a shirt for £35 and a pair of shoes for £45. How much money did she spend in total?

___

10 At a charity event, 72 people attend in the morning and 95 people attend in the afternoon. How many people attended the event in total?

___

11 Maria goes shopping for groceries. She spends £124 on fruits and vegetables, £78 on dairy products, and £39 on baking supplies. Calculate the total amount she spends.

____

Progress check

1 Dan is buying a laptop and a TV.

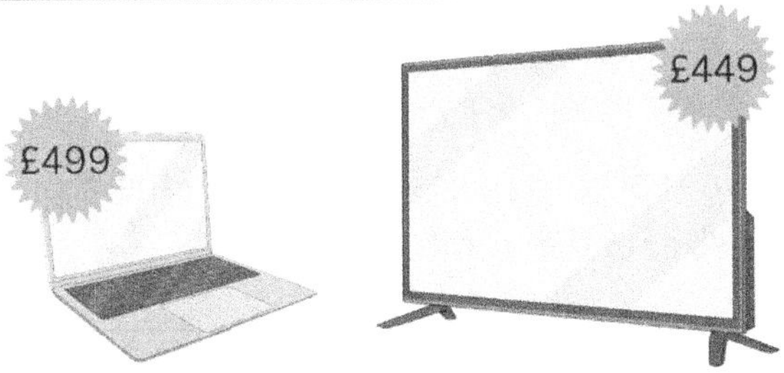

How much do both items cost in total?

____

____

2 How many cubes are there altogether? Show your working.

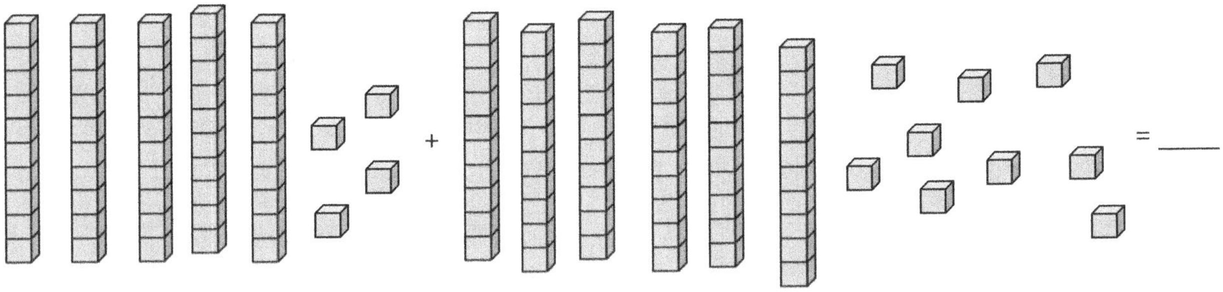

____

3 Fill in the gaps below:

527 + 115 = _____	120 + _____ = 535	_____ + 99 = 399
404 + 286 = _____	658 + _____ = 757	_____ + 68 = 503

4 Naila writes these maths sentences:

249 + 154 = 154 + 249

0 + 135 = 0

Do you agree with Naila? Explain your answer.

____

5 Find another two numbers you can add which give the same total as 325 + 246

____

6 In a gym, 43 people are using the treadmills and 136 people are using the exercise bikes. How many people are exercising in total?

____

Subtract three-digit numbers

The part-whole method of subtraction

This is one method you can use to subtract.

35 - 23

We split 35 into 30 and 5.
We split 23 into 20 and 3.

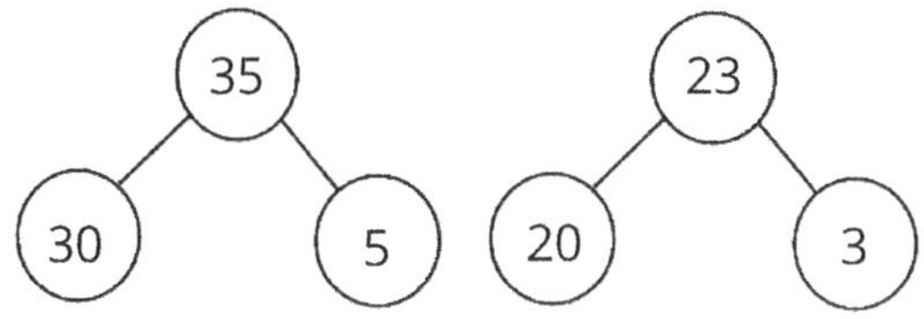

Now we subtract the tens: 30 - 20 = 10
And the ones: 5 - 3 = 2
Altogether: 10 + 2 = 12

257 - 124

We split 253 into 200 and 50 and 7.
We split 123 into 100 and 20 and 4.

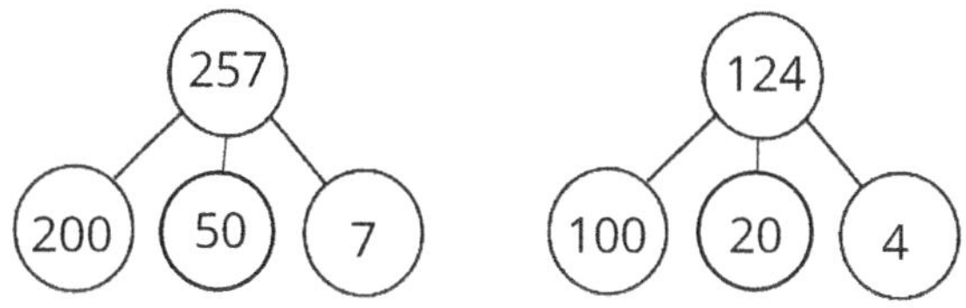

Now we subtract the hundreds: 200 - 100 = 100
And the tens: 50 - 20 = 30
And the ones: 7 - 4 = 3
Altogether: 100 + 30 + 3 = 133

1 Fill the gaps.

976 - 302

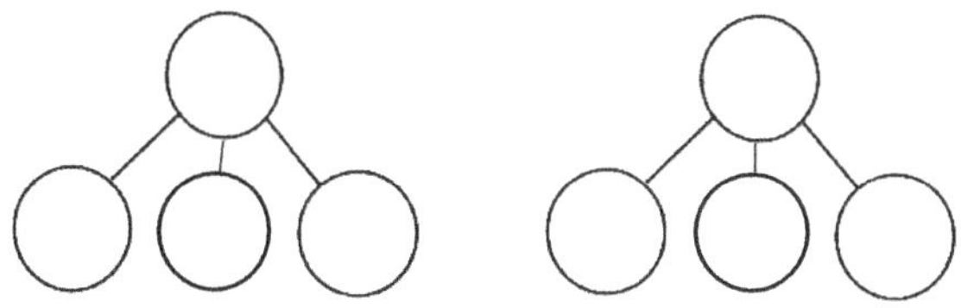

573 - 260

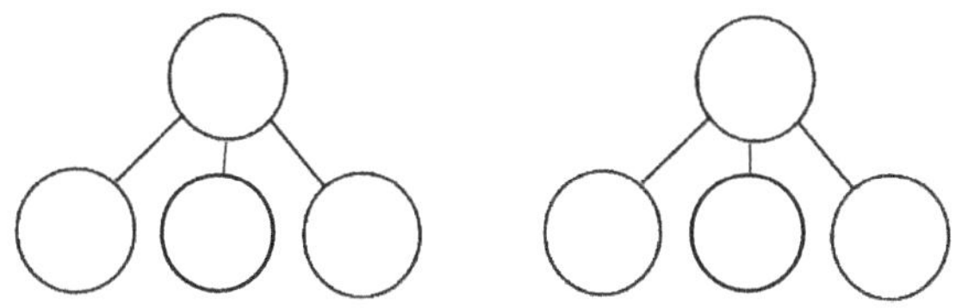

328 - 305

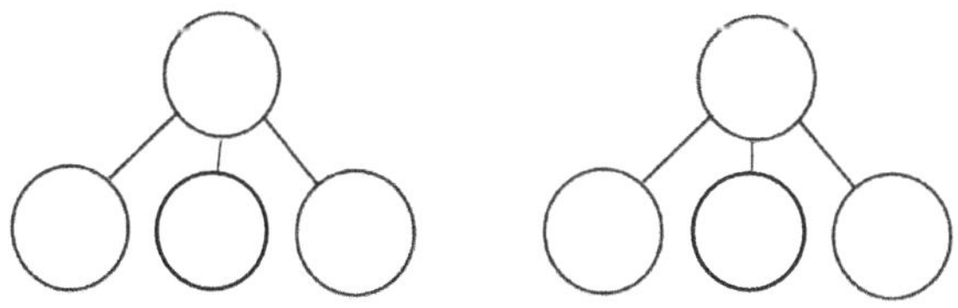

569 - 237

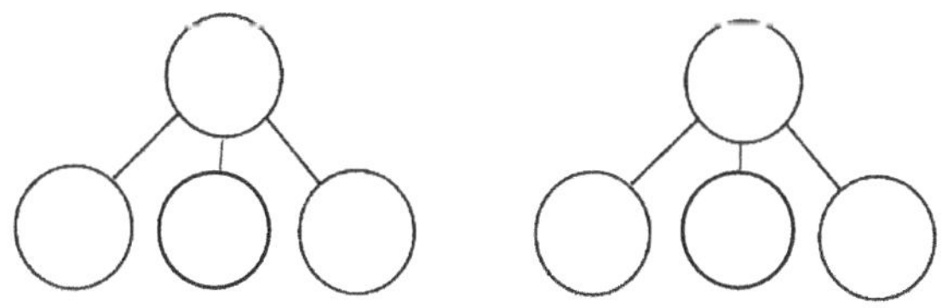

482 - 332

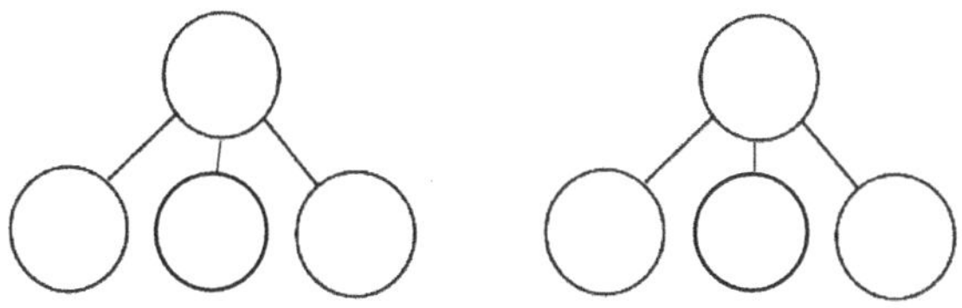

463 - 235

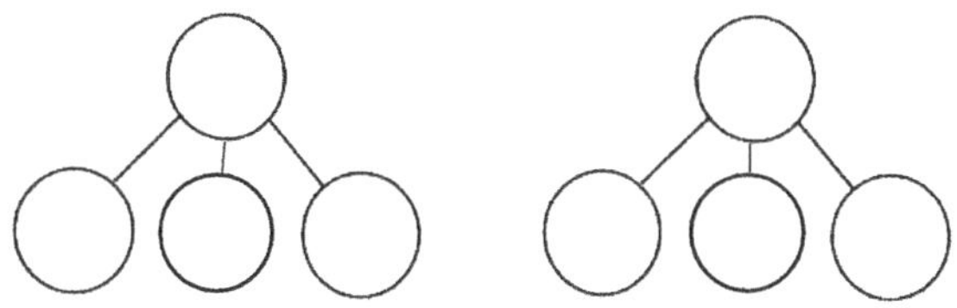

2 Fill the gaps.

651 - 540

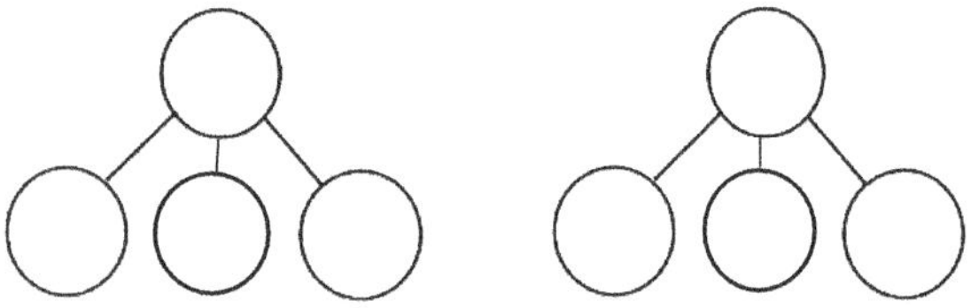

392 - 242

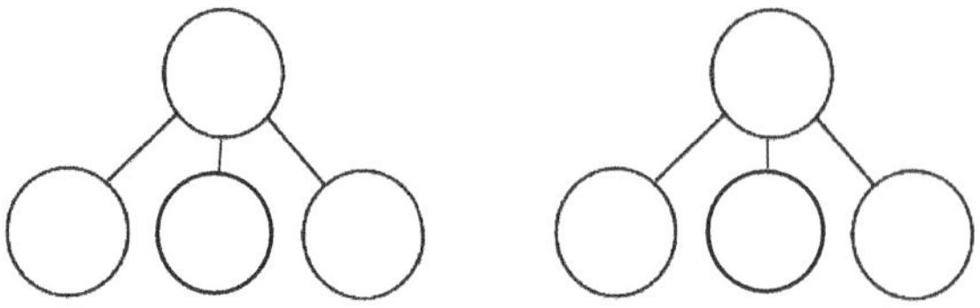

629 - 206

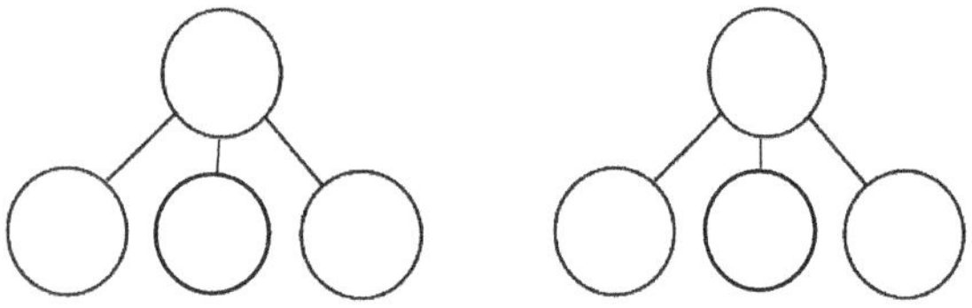

534 - 313

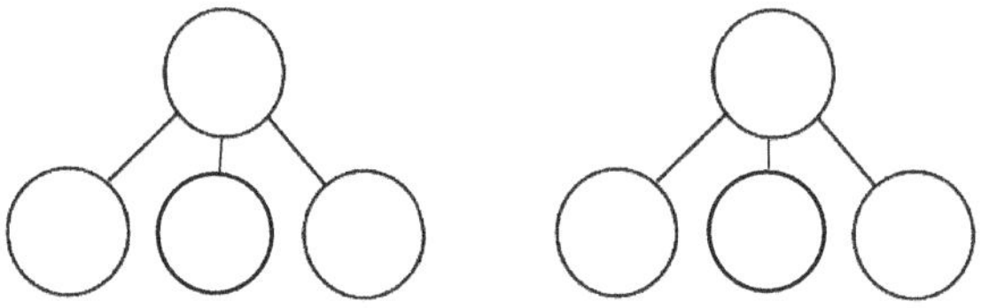

305 - 205

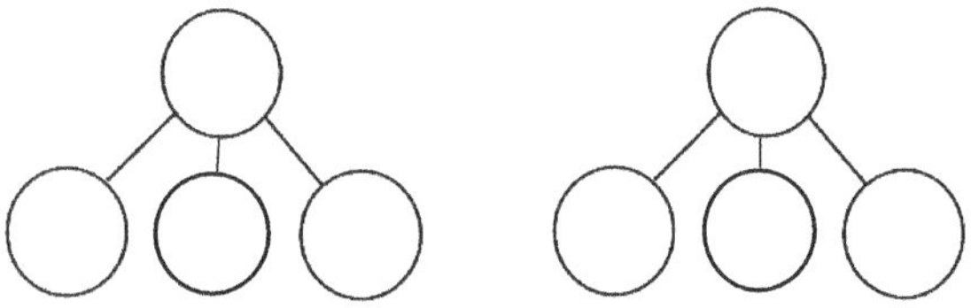

754 - 642

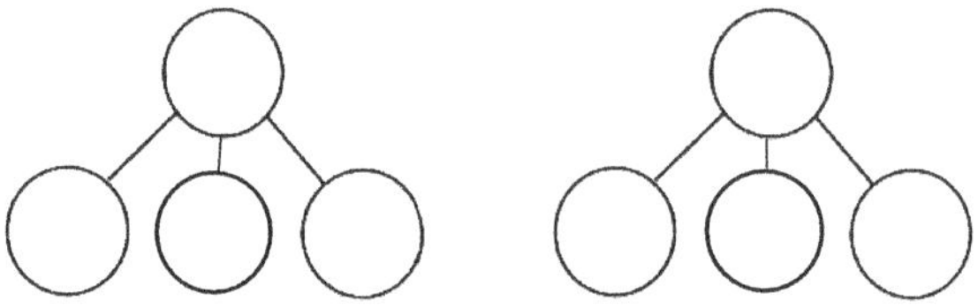

353 - 202

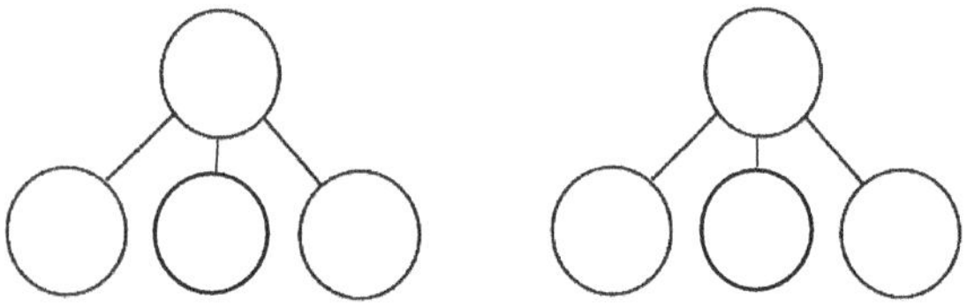

469 - 226

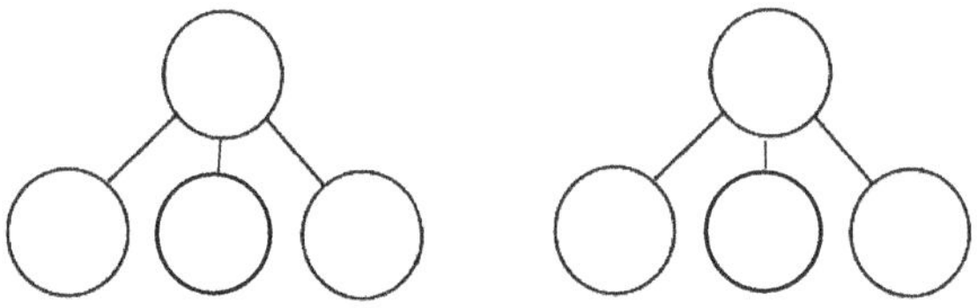

463 - 232

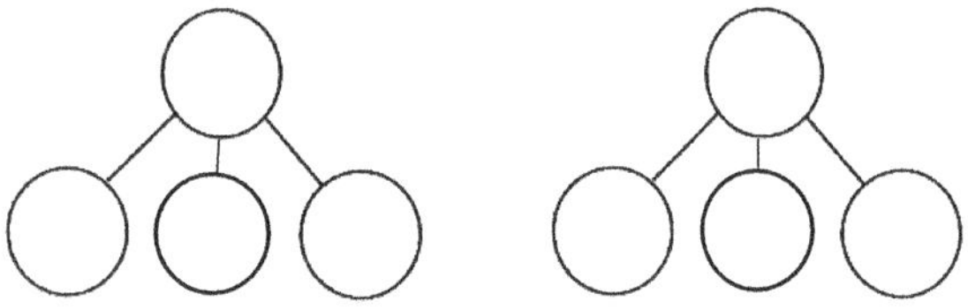

976 - 38

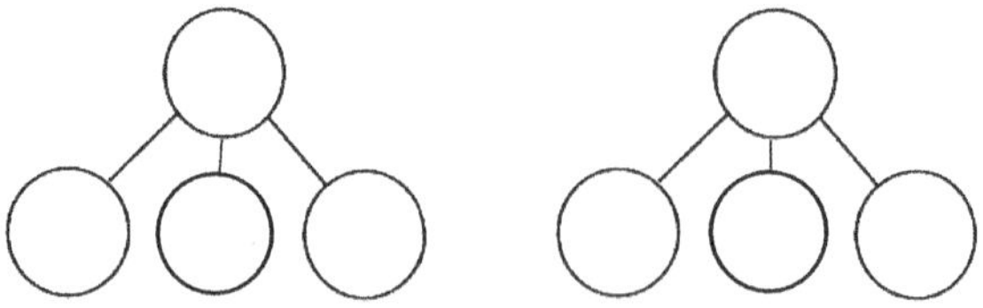

The column method of subtraction

Example 1

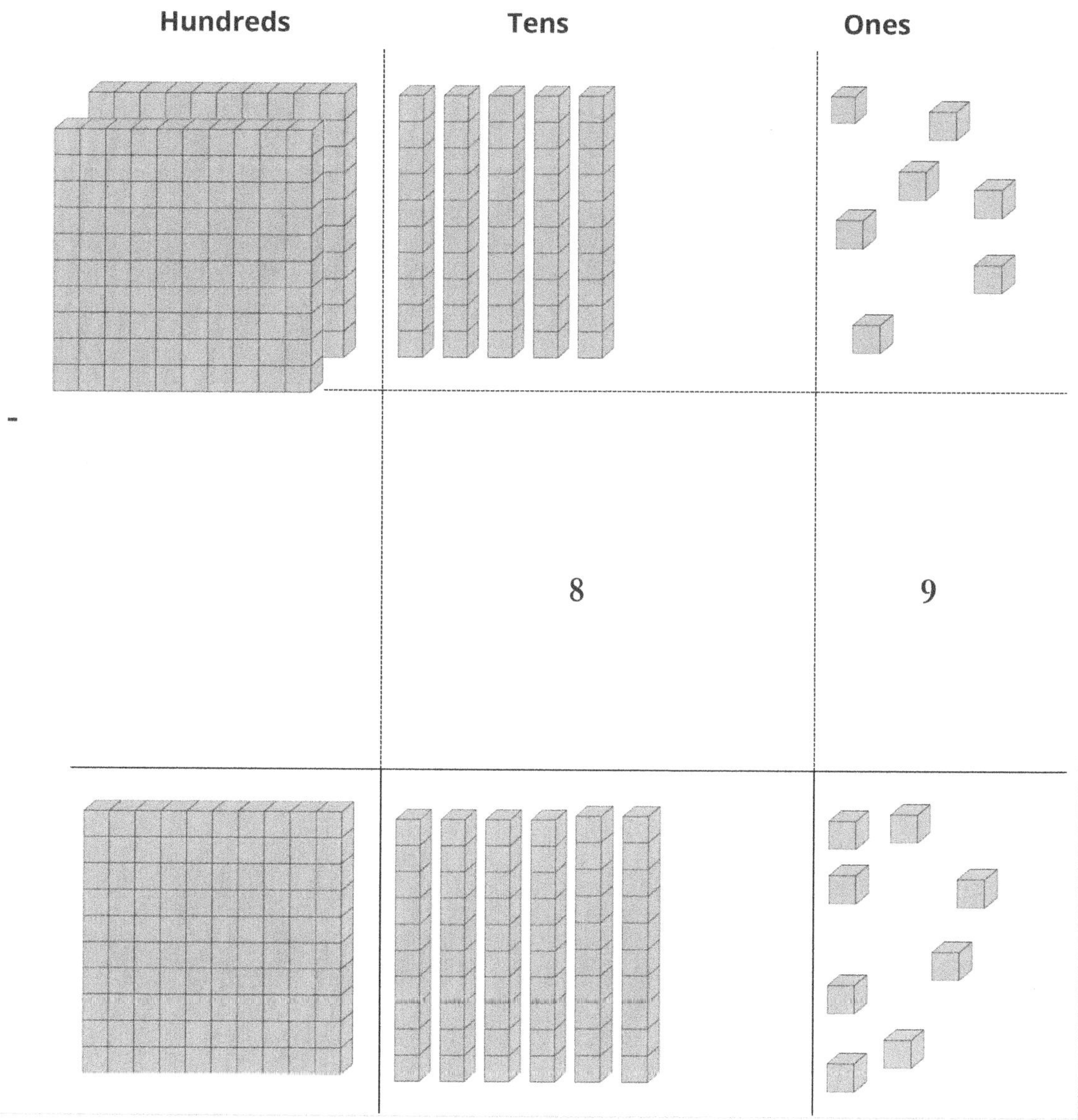

Hundreds	Tens	Ones
1	4	1
~~2~~	~~5~~	7
-	8	9
1	6	8

As we can't take away 9 from 7, we borrow one group of 10. So, 7 becomes 17. Therefore, 17 – 9 = 8

There are 4 groups of 10 left after the borrowing. We need to take away 8 from the 4, so we borrow a group of 100, which makes 14 tens.

14 – 8 = 6. Then 1 – 0 = 1

Example 2

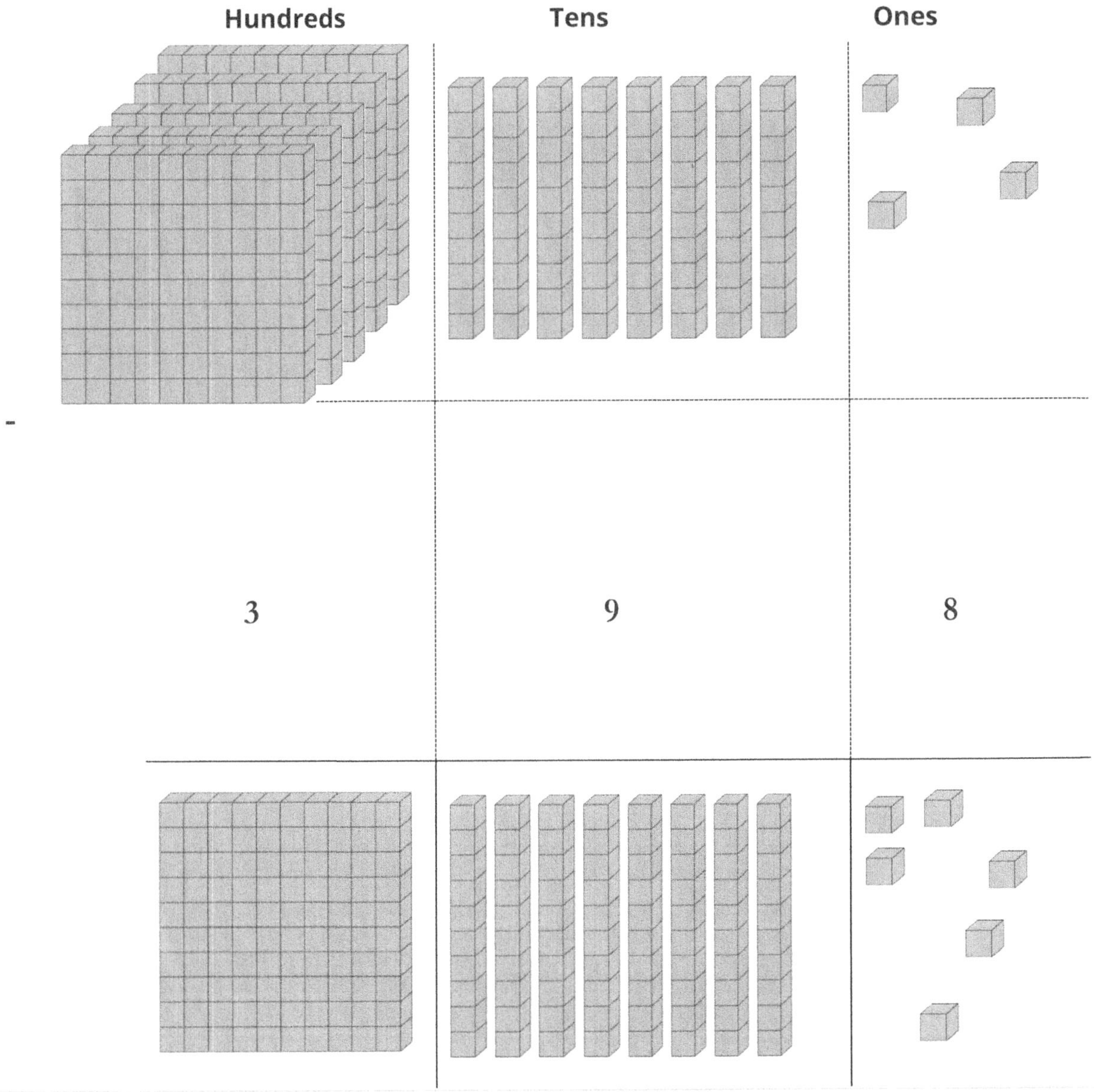

Hundreds	Tens	Ones
4	7	1
~~5~~	~~8~~	4
- 3	9	8
1	8	6

As we can't take away 8 from 4, we borrow one group of 10. So, 4 becomes 14. Therefore, 14 – 8 = 6

There are 7 groups of 10 left after the borrowing. We need to take away 9 from the 7, so we borrow a group of 100, which makes 17 tens.

17 – 9 = 8. Then 4 – 3 = 1

Here are some more examples.

$\overset{2}{\cancel{3}}$	$\overset{7}{\cancel{8}}$	$^{1}5$	$\overset{6}{\cancel{7}}$	$\overset{9}{\cancel{^{1}0}}$	$^{1}2$	$\overset{3}{\cancel{4}}$	$\overset{9}{\cancel{^{1}0}}$	$^{1}0$
- 2	8	7	-	9	9	- 3	7	9
	9	8	6	0	3		2	1

When there are 0 tens, we borrow from the hundreds. Then there are 10 tens you can borrow from.

3 Fill the gaps.

191 - 67 = ___	424 - 185 = ___	756 - 84 = ___	690 - 387 = ___	793 - 73 = ___	913 - 96 = ___	814 - 157 = ___
508 - 481 = ___	576 - 327 = ___	415 - 372 = ___	484 - 226 = ___	126 - 77 = ___	791 - 538 = ___	664 - 74 = ___
641 - 348 = ___	870 - 458 = ___	954 - 102 = ___	572 - 427 = ___	765 - 174 = ___	792 - 109 = ___	553 - 283 = ___
695 - 205 = ___	831 - 175 = ___	362 - 276 = ___	559 - 310 = ___	814 - 327 = ___	386 - 157 = ___	458 - 396 = ___
642 - 285 = ___	273 - 135 = ___	516 - 473 = ___	978 - ___ = 389	768 - ___ = 332	674 - ___ = 578	580 - ___ = 485

You can use whichever method you like to subtract. It is good practice to know at least two methods as it helps with deeper understanding and in case you forget one under exam conditions.

Use a calculator to subtract these numbers.

56 – 14 = ____ and 14 – 56 = _____

100 – 63 = ____ and 63 – 100 = _____

200 – 134 = _____ and 134 – 200 = _____

What do you notice?

Order matters when subtracting.
74 - 43 = 31   43 - 74 = -31

4 A phone normally costs £699. The store is offering a discount of £120 off the total cost. Work out the discounted price of the phone.

___

5 Carlos has £550 in his bank account. He withdraws £95 for a new pair of shoes and £150 for a concert ticket. How much money does he have left in his bank account?

___

6 The local gym had 180 visitors on Monday and 205 visitors on Tuesday. How many more visitors did the gym have on Tuesday than on Monday?

___

7 Donna has £153. She spends £115 on food.
Donna wants to work out how much money she has left: 153 – 115 = 38
Is Donna correct?

___

Language is important!

Subtract 7 **from** 9:
Example: 9 – 7 = 2

8 During a charity event, Emily raised £245 from a bake sale and £330 from a sponsored run. Emily works out how much more she raised from the sponsored run than the bake sale: 245 - 330 = 85
Is Emily correct?

___

9 A charity collected £580 in donations this month. They spent £410 on supplies. To work out how much money is left, the manager calculates: 580 + 410 = 990. Are they correct?

___

10 Nicole says: 'When subtracting 0 from a number, the number stays the same'. Is Nicole correct?

___

The link between addition and subtraction

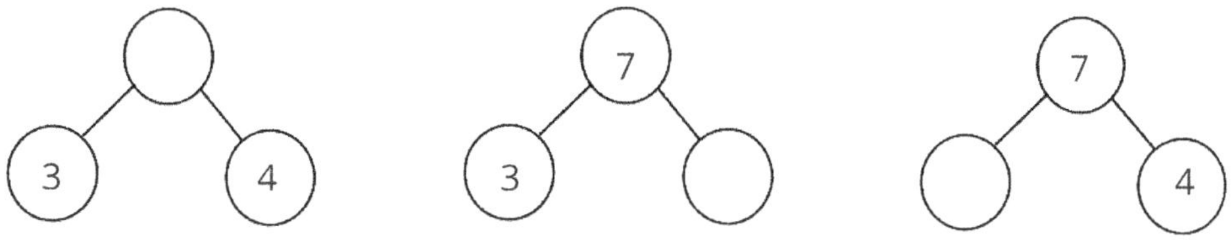

3 + 4 = 7 7 – 3 = 4 7 – 4 = 3

What do you notice?

3 + ___ = 7 is the same as 7 – 3

___ + 4 = 7 is the same as 7 – 4

Addition ⟷ Subtraction
Addition is the opposite of subtraction.
Subtraction is the opposite of addition.

11 Fill the gaps. The first one has been done for you.

86 + ___ = 113 is the same as 113 – 86	___ + 657 = 913 is the same as 913 – 657
239 + ___ = 414 is the same as _________	___ + 345 =414 is the same as _________
605 + ___ = 712 is the same as _________	___ + 877 = 912 is the same as _________
248 + ___ = 519 is the same as _________	___ + 511 = 719 is the same as _________

12 Fill the gaps.

270 take away 98 is written as 270 – 98

330 take away 109 is written as ______

590 take away 258 is written as ______

600 take away 161 is written as ______

464 take away 384 is written as ______

Take away 164 **from** 247 is written as 247 – 164

Take away 362 from 458 is written as ______

Take away 245 from 317 is written as _______

Take away 429 from 484 is written as _______

Take away 178 from 859 is written as ______

Checking answers

Example: Naila spends £60 on a new shirt and £35 on a pair of jeans. To calculate the total amount she spends, we do: 60 + 35 = 95 **To check the answer:** 95 – 35 = 60 or 95 – 60 = 35

Example: A video game costs £110. The shop is offering a discount of £19 off the total cost. To work out the discounted price of the video game, we do: 110 - 19 = 91 **To check:** 91 + 19 = 110

Example: A restaurant served 276 meals on Saturday and 312 meals on Sunday. To work out how many meals were served in total over the weekend, we do: 276 + 312 = 588
To check: 588 - 312 = 276 or 588 - 276 = 312

Example: Olivia has £375 in her savings account. She withdraws £279 for a new tablet. To work out how much money she has left, we do: 375 – 279 = 96 **To check**: 96 + 279 = 375

13 A bicycle costs £499. The shop is offering a discount of £150 off the total cost. How much will the bicycle cost with this offer?

____

Check your answer: ______________________________

14 At a sports event, Team A and Team B are playing. There are 219 Team A fans and 310 Team B fans attending. How many fans are attending in total?

____

Check your answer: ______________________________

15 At a book fair, there were 295 attendees from School A and 250 attendees from School B. How many attendees were there in total from both schools?

____

Check your answer: ______________________________

16 Neil has £489 in his savings account. He withdraws £35 for a new book and £80 for a concert ticket. How much money does he have left in his savings account?

____

Check your answer: ______________________________

17 Kim spends £135 on groceries and £22 on toiletries. Calculate the total amount she spends.

____

Check your answer: ______________________________

18 A music festival recorded the number of attendees over three days:
Friday - 300 attendees, Saturday - 70 attendees, Sunday - 250 attendees. How many attendees were there in total during the festival?

____

Check your answer: ______________________________

> **Difference** means **taking away**.
> Example: The difference between 9 and 7 is 2.

19 a) What is the difference between 230 and 115? ____________________

b) Work out the difference between 400 and 326. ____________________

c) What is the difference between 732 and 584? ____________________

d) Work out the difference between 960 and 495. ____________________

20

a) A school has 307 students, and each student needs a laptop for digital learning. The school has already purchased 125 laptops.

How many more laptops does the school need to ensure every student has one?

b) Sam has a digital file that is 250MB in size. He wants to upload it to his cloud storage account, but he only has 175MB of free space left.

How much more space does he need to upload the file?

c) Zen is travelling to France. The distance from his town to the city he is going is 392 miles.

What is the total distance Zen will travel to go there and back?

d) Roy's rent is £525 per month.
From next month, the rent will go up by £16.

How much will the rent be after the increase?

£

e) Soni is going on holiday. She needs 150ml of her medication while on holiday. She already has 85ml. How much more does she need to buy?

f) In a mobile game, Emily has collected 350 gold coins and 200 silver coins.

How many coins has Emily collected in total?

g) A diversity scholarship fund has £800 available for scholarships.

If £550 has already been awarded, how much money is left in the fund for future scholarships?

h) A weight loss program suggests burning 500 calories per workout session. If Alice has already burned 350 calories, how many more calories does she need to burn to meet the program's suggestion?

i) During a cultural festival, 135 participants are wearing traditional outfits and 227 are wearing casual clothes.

How many participants are there in total?

j) Joshua's phone has a 256GB memory capacity. He adds memory card with 64GB to the phone.
How many GB does the phone have, including the memory card?

k) A company employs 350 people. It aims to have at least 200 employees from diverse backgrounds. Currently, they employ 120 people from these backgrounds.

How many more of these employees does the company need to hire to meet its goal?

l) Peter has saved 500 game credits. He spends 200 credits on a new character and 100 credits on an equipment upgrade.

How many game credits does Peter have left after these purchases?

21 Fill the gaps with numbers. The first one has been done for you.

118	+	256	=	374
	+		=	532
	+		=	641
343	=		+	
267	=		+	

306	-	177	=	129
	-		=	215
	-		=	466
318	=		-	
127	=		-	

22 Fill the gaps. The first one has been done for you.

284	+	167	=	451
	+		=	451
	+		=	451
451	=		+	
451	=		+	

538	-	372	=	166
	-		=	166
	-		=	166
166	=		-	
166	=		-	

Subtract these numbers.

54 - 54 = ____ 130 - 130 = ____ 166 - 166 = ____ 609 - 609 = ____

115 - 115 = ____ 107 - 107 = ____ 571 - 571 = ____ 144 - 144 = ____

What do you notice?

When you subtract the number itself, you get 0.
12 - 12 = **0** 13 - 13 = **0**

23 Work out:

250 - 250 =	186 – 186 =	567 – 567 =	121 – 121 =	205 - 205 =	0 - 0 =

24 Fill the gaps.

	3	6	8
+	4	2	
	7	9	5

	6		8
+	2	3	5
	8	6	

	8		4
+		4	
	8	8	2

	4	2	3
+	1		
		6	1

25 Fill the gaps.

	5	6	8
-		2	
	1	4	5

	6		8
-	2	3	
	4	4	9

	3		4
-	1	4	
		1	7

	8	7	3
-			
	3	6	9

26 Ayla is adding 62 and 18. This is how she has lined the numbers up.
What has she done wrong?

____

		6	2
+	1	8	
	2	4	2

Mental Methods: Number bonds

When adding and subtracting, it is very useful to know the number bonds. Number bonds are pairs of numbers that total 5 or any other number.

These are the pairs of numbers that make 5. These are called **number bonds to 5**.

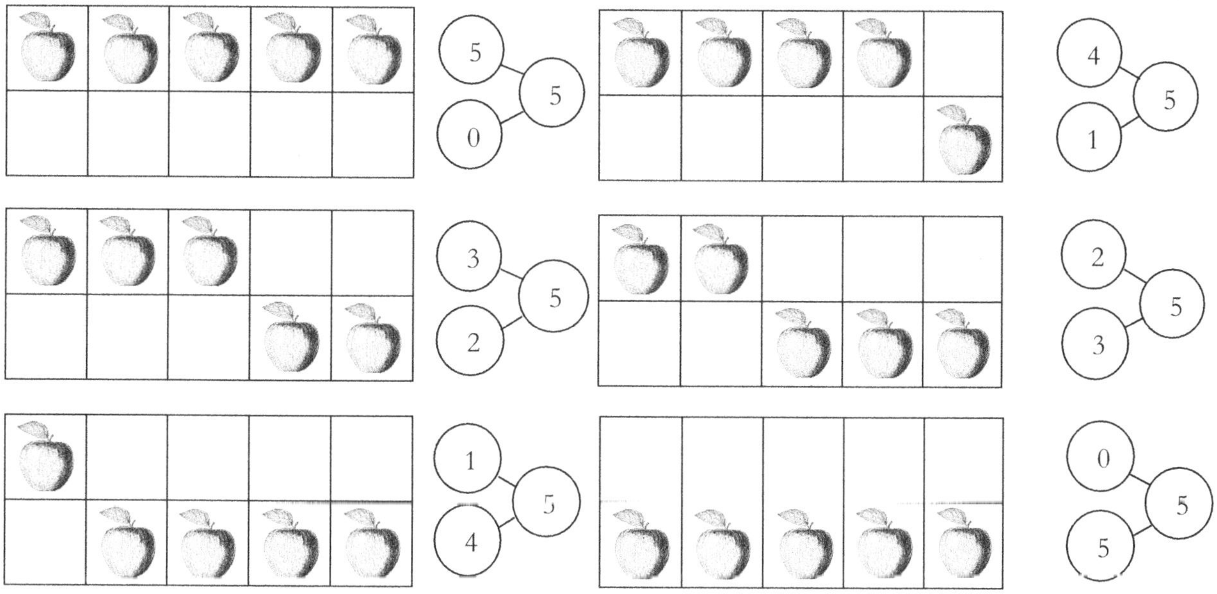

27 Fill these gaps.

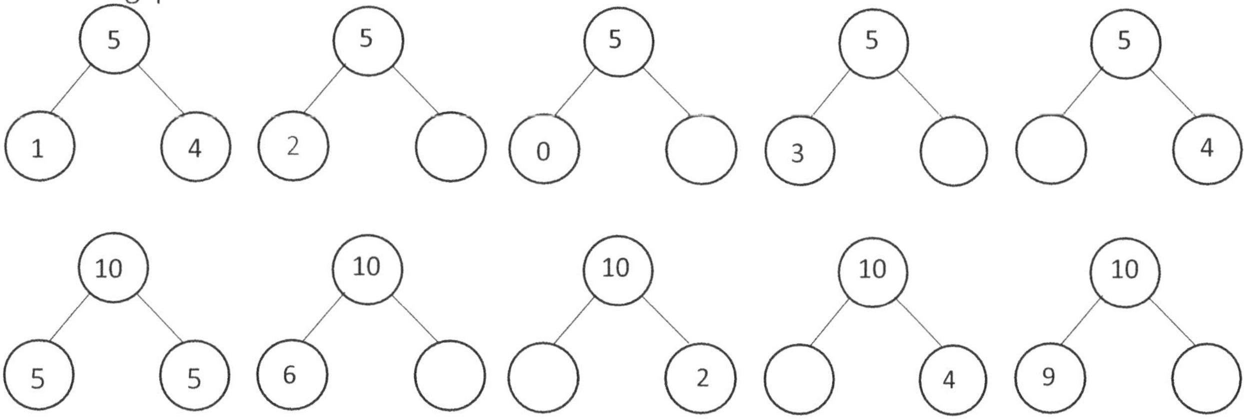

28 Fill these gaps:

1 + ___ = 5	2 + ___ = 5	___ + 4 = 5	___ + 5 = 5	___ + [illegible] = 5
3 + ___ = 5	4 + ___ = 5	___ + 2 = 5	3 + 2 = ___	___ + 1 = 5
0 + ___ = 10	1 + ___ = 10	___ + 7 = 10	___ + 0 = 10	2 + ___ = 10
3 + ___ = 10	4 + ___ = 10	___ + 6 = 10	___ + 2 = 10	5 + ___ = 10
6 + ___ = 10	7 + ___ = 10	___ + 4 = 10	___ + 9 = 10	8 + ___ = 10

Mental methods: Compensation

When adding 53 and 9:
As 9 is close to 10, add 10 and take away 1.

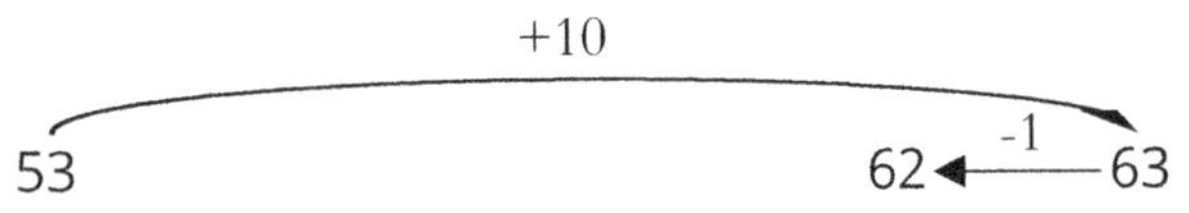

So, 53 + 9 = 62

When subtracting 19 from 72:
As 19 is close to 20, subtract 20 and add 1.

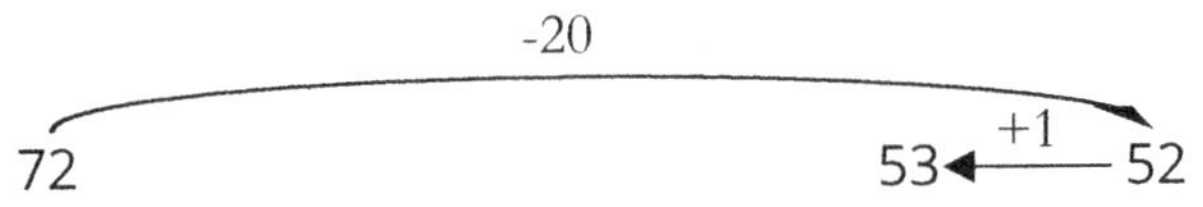

So, 72 - 19 = 53

When adding 68 and 28:
As 28 is close to 30, add 30 and take away 2.

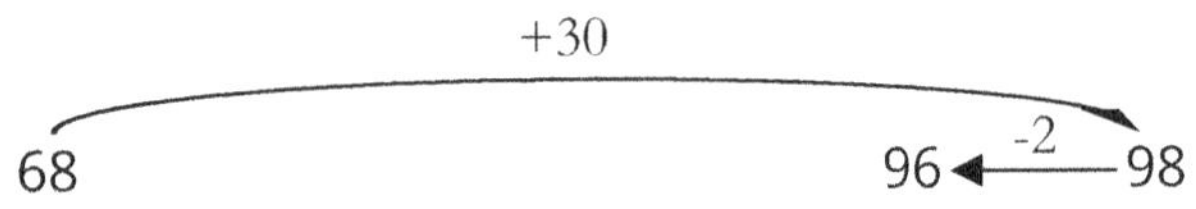

So, 68 + 28 = 96

When subtracting 37 from 95:
As 37 is close to 40, subtract 40 and add 3.

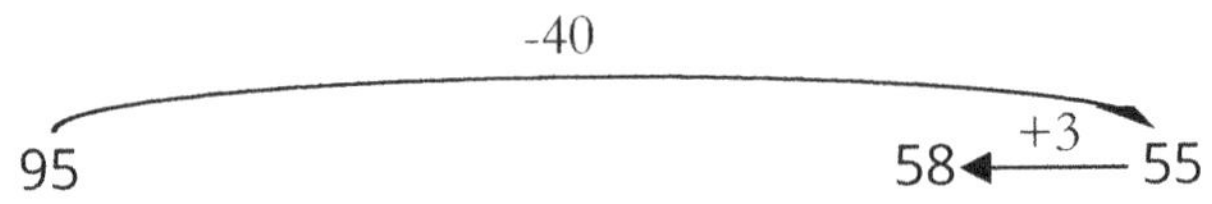

So, 95 - 37 = 58

Fill the gaps.

29 35 + 9 = ___

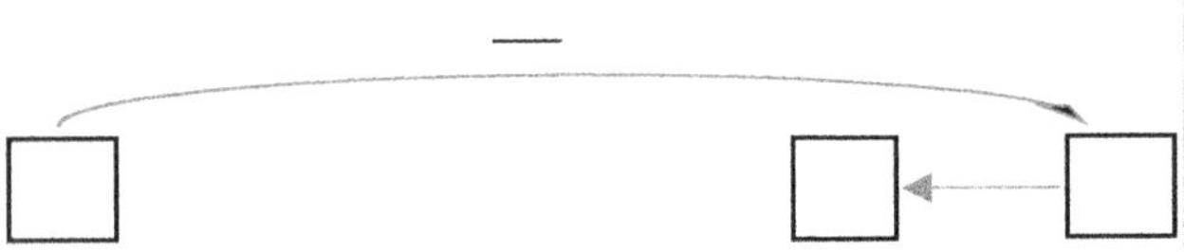

42 - 18 = ___

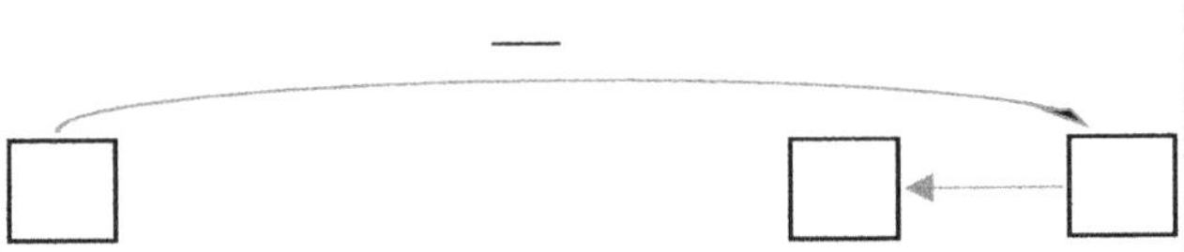

56 + 27 = ___

77 - 9 = ___

77 + 28 = ___

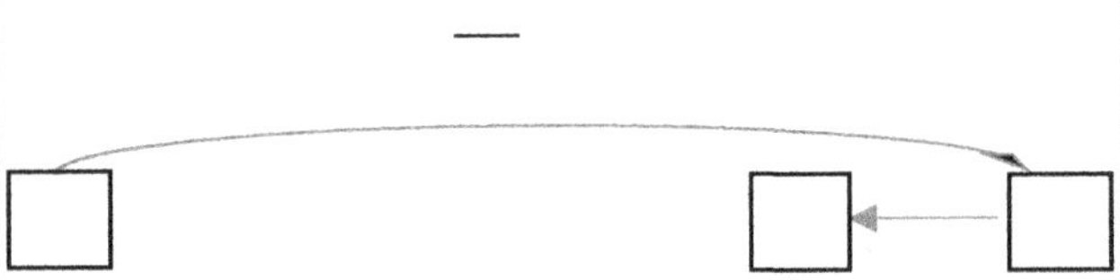

56 - 29 = ___

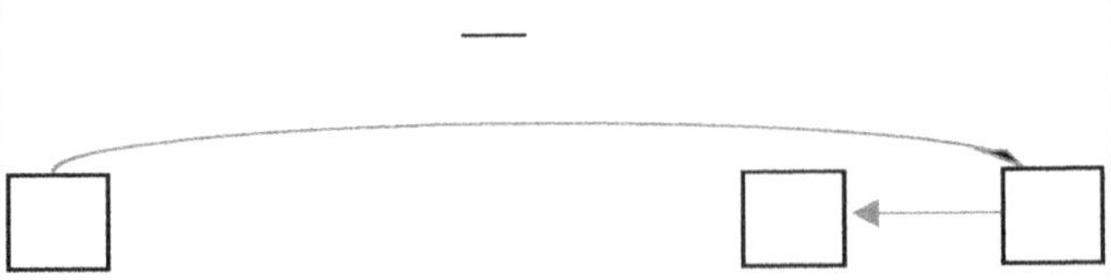

Mental methods: Reordering

When adding: 6 + 8 + 4	When taking away: 17 – 4 – 7
6 + 4 = 10 (number bonds)	17 - 7 = 10
10 + 8 = 18	10 - 4 = 6 (number bonds)

30 Fill the gaps.

5 + 9 + 5 =	25 - 6 - 5 =
□ + □ = □	□ - □ = □
□ + □ = □	□ - □ = □
11 + 7 + 9 =	**38 - 3 - 8 =**
□ + □ = □	□ - □ = □
□ + □ = □	□ - □ = □

Progress check

1 Dan is buying a laptop and a TV.
He has a £1000 budget. Is he within the budget?

____

____

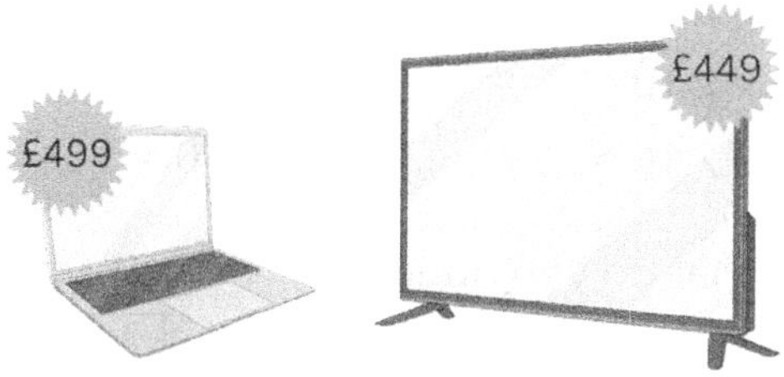

2 How many cubes are left after the subtraction? Show your working.

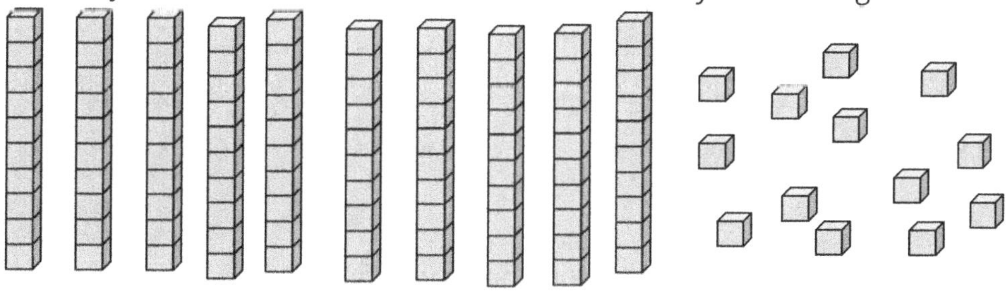

- 86 = _____

3 Fill in the gaps below:

439 - 215 = _____	755 - _____ = 612	_____ - 459 = 499
544 - 326 = _____	876 - _____ = 268	_____ - 4 = 801

4 Naila writes these maths sentences:

124 - 4 = 4 - 124

250 - 0 = 250

Do you agree with Naila? Explain your answer.

____

Multiply two-digit whole numbers by single and double-digit whole numbers

We multiply when calculating:

- the cost of multiple items
- how long it takes to do a task again and again
- pay per week, given number of hours worked
- how much you will pay per month or year for bills
- how much of each ingredient you need when cooking
- how much paint you may need to paint walls

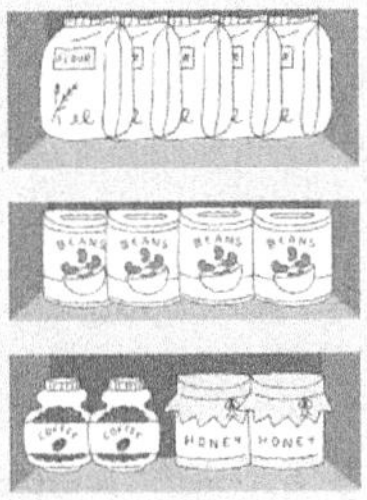

Multiplying by 10

This chocolate bar has 5 rows with 10 squares in each row.

To work out the total number of squares, we can do:
10 + 10 + 10 + 10 + 10
which is the same as 5 lots of 10.

Therefore: **5 x 10 = 50**

Use a calculator to work out:

6 x 10 =	11 x 10 =	75 x 10 =	10 x 10 ≈	9 x 10 =
7 x 10 =	12 x 10 =	81 x 10 =	43 x 10 ≈	10 x 89 =
8 x 10 =	13 x 10 =	10 x 48 =	10 x 37 ≈	26 x 10 =

What do you notice?

When multiplying a whole number by 10, you just add a 0 at the end.

1 Work out without a calculator.

6 x 10 =	11 x 10 =	75 x 10 =	10 x 10 ≈	9 x 10 =
7 x 10 =	12 x 10 =	81 x 10 =	43 x 10 ≈	10 x 89 =
8 x 10 =	13 x 10 =	10 x 48 =	10 x 37 ≈	26 x 10 =

2 If you work 10 hours a day, how many hours do you work in two weeks?

___

3 Dan is buying ten tins of soup. One tin costs 45p. How much will ten tins cost in total? Give your answer in pence.

___

4 Ali has ten vintage books. She wants to sell them at £15 each. How much will all ten books be worth?

___

Multiplying two-digit by one-digit numbers

This chocolate bar has 5 rows with 13 squares in each row.

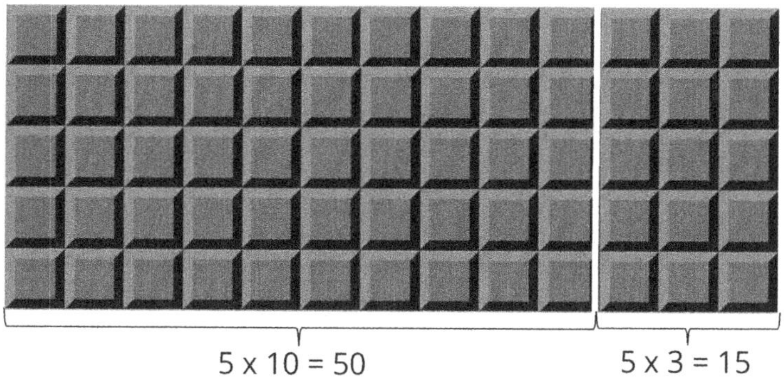

To work out 5 x 13, you can split 13 into **10** and **3**

Then multiply
5 x 10 = 50
5 x 3 = 15

Finally: 50 + 15 = 45

Here are some more examples.

Example: 14 x 7 ➡ 10 x 7 = 70 and 4 x 7 = 28, so 70 + 28 = 98

Example: 15 x 6 ➡ 10 x 6 = 60 and 5 x 6 = 30, so 60 + 30 = 90

Example: 18 x 3 ➡ 10 x 3 = 30 and 8 x 3 = 24, so 30 + 24 = 54

Example: 17 x 9 ➡ 10 x 9 = 90 and 7 x 9 = 63, so 90 + 63 = 153

5 Work out without a calculator.

6 x 17 =	9 x 15 =	11 x 9 =	13 x 4 =	14 x 8 =
7 x 13 =	5 x 16 =	14 x 7 =	15 x 5 =	16 x 9 =
8 x 12 =	6 x 15 =	17 x 8 =	11 x 8 =	15 x 7 =

6 A tea bag costs 18p. How much do four tea bags cost?

7 There are 6 eggs in one carton. How many eggs are there in 15 cartons?

8 Use a calculator to work out:

6 x 20 =	9 x 30 =	9 x 20 =	40 x 80 =	50 x 80 =
7 x 40 =	80 x 5 =	3 x 70 =	70 x 30 =	80 x 80 =
8 x 50 =	3 x 30 =	60 x 60 =	80 x 40 =	90 x 30 =

What do you notice?

When multiplying a number by a multiple of 10 (like 20, 30, 40, etc.), first multiply the number by the first digit of the multiplier and then add as many zeros to the result as there are in the multiplier.

Example: 7**0** x 8**0** = 56**00**

9 Work out without a calculator.

6 x 40 =	9 x 20 =	9 x 80 =	40 x 50 =	60 x 80 =
7 x 60 =	70 x 5 =	30 x 20 =	70 x 70 =	30 x 80 =
8 x 30 =	3 x 50 =	60 x 40 =	80 x 50 =	80 x 30 =

This chocolate bar has 5 rows with 21 squares in each row. To work out 5 x 21, you can split 21 into 20 and 1.

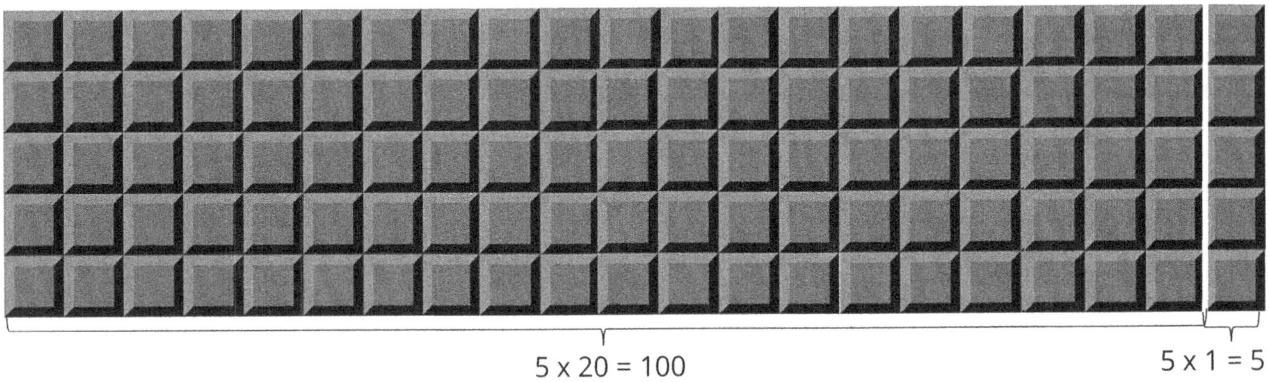

Then 100 + 5 = 105

Here are some more examples.

Example: 35 x 7 ➡ 30 x 7 = 210 and 5 x 7 = 35, so 210 + 35 = 245

Example: 52 x 6 ➡ 50 x 6 = 300 and 2 x 6 = 12, so 300 + 12 = 312

Example: 68 x 3 ➡ 60 x 3 = 180 and 8 x 3 = 24, so 180 + 24 = 204

Example: 84 x 9 ➡ 80 x 9 = 720 and 4 x 9 = 36, so 720 + 36 = 756

10 Work out without a calculator.

6 x 37 =	6 x 85 =	55 x 5 =	9 x 35 =	43 x 4 =
7 x 53 =	81 x 9 =	21 x 8 =	5 x 46 =	96 x 9 =
8 x 62 =	74 x 7 =	84 x 8 =	57 x 8 =	85 x 7 =

11 A can of tuna weighs 85g. How much do 6 cans of tuna weigh? Give your answer in grams.

____

12 A book club subscription costs £18 per month. If Emily wants to buy a 6-month subscription, how much will it cost her in total?

____

13 A can of soup costs 54p. How much will 7 cans cost? Give your answer in pence.

____

Multiplying two-digit by two-digit numbers

This chocolate bar has 15 rows with 13 squares in each row.

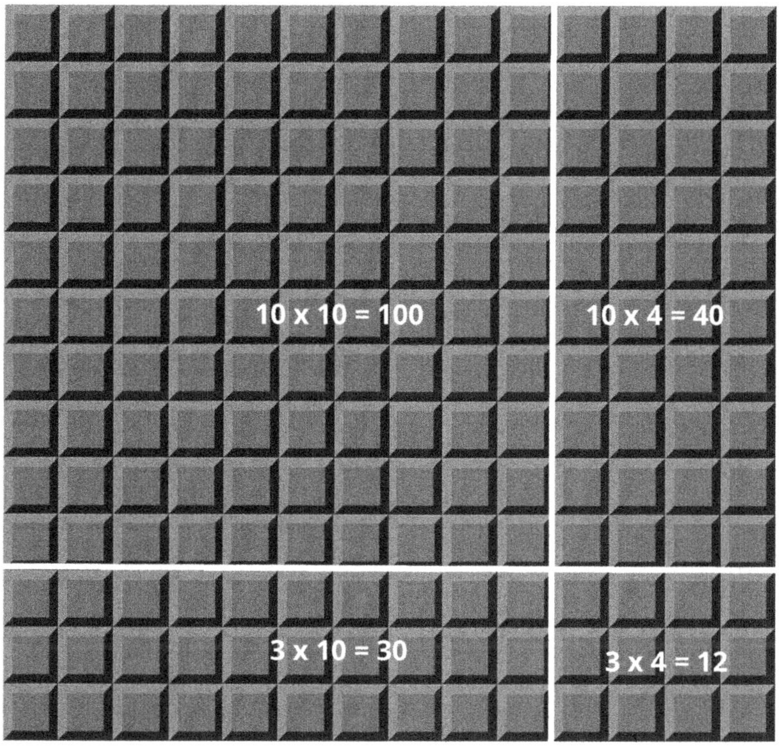

To work out 13 x 14, you can split:

13 into **10** and **3**
14 into **10** and **4**

x	**10**	**4**
10	100	40
3	30	12

Then multiply:
10 x 10 = 100
10 x 4 = 40
3 x 10 = 30
4 x 3 = 12

Finally: 100 + 40 + 30 + 12 = 182

Here are some more examples.

Example: 24 x 36

x	**20**	**4**
30	600	120
6	120	24

600 + 120 + 120 + 24 = 864

Example: 25 x 37

x	**30**	**7**
20	600	140
5	150	35

600 + 150 + 140 + 35 = 925

Example: 43 x 17

x	**10**	**7**
40	400	280
3	30	21

400 + 280 + 30 + 21 =731

Example: 29 x 24

x	**20**	**9**
20	400	180
4	80	36

400 + 180 +80 + 36 = 396

Example: 43 x 17

x	**10**	**7**
40	400	280
3	30	21

400 + 280 + 30 + 21 =731

Example: 29 x 24

x	**20**	**9**
20	400	180
4	80	36

400 + 180 +80 + 36 = 396

14 Multiply these numbers.

37 x 14

x		

28 x 24

x		

44 x 21

x		

39 x 23

x		

54 x 15

x		

38 x 16

x		

52 x 13

x		

45 x 22

x		

62 x 15

x		

13 x 72

x		

The column method of multiplication

Example 1

	1	
	5	4
x		3
1	6	2

54 x 3

We start with the units so that if we have to carry over, we can do so.

3 x 4 = 12, so we put 2 down and carry over the tens (1 in this case).

3 x 5 = 15 Add to that 1 that we carried over and that makes 16. Write 16 down.

So, 54 x 3 = 162

Example 2

	3	
	6	5
x		7
4	5	5

65 x 7

We start with the units so that if we have to carry over, we can do so.

7 x 5 = 35, so we put 5 down and carry over the tens (3 in this case).

7 x 6 = 42 Add to that 3 that we carried over and that makes 45. Write 45 down.

So, 65 x 7 = 455

Example 3

	7	
	7	8
x		9
7	0	2

78 x 9

We start with the units so that if we have to carry over, we can do so.

9 x 8 = 72, so we put 2 down and carry over the tens (7 in this case).

9 x 7 = 63 Add to that 7 that we carried over and that makes 70. Write 70 down.

So, 78 x 9 = 702

15 Work out without a calculator.

43×8	72×9	27×5	56×6
81×4	57×3	92×5	73×6
64×7	89×4	63×3	94×8
72×6	38×5	62×8	67×9

Example 4

		1	
		5	4
x		1	3
	1	6	2
+	5	4	0
	7	0	2

We start with the units so that if we have to carry over, we can do so.

3 x 4 = 12, so we put 2 down and carry over the tens (1 in this case).

3 x 5 = 15 Add to that 1 that we carried over and that makes 16. Write 16 down.

Now, we finished multiplying 3 by the top numbers. So will multiply 1 by the top numbers, always starting with units (4 in this case).

The value of the digit 1 is 10. You know that multiplying a whole number by 10, is like adding a 0 at the end. Rather than doing 10 x 4 and then 10 x 5, we just put a 0 down in the second row of answers.

Then 1 x 4 = 4 and 1 x 5 = 5

Adding the two answers: 162 + 540 = 702 So, 54 x 13 = 702

Example 5

		3	
		3	5
x		2	7
	2	4	5
+	7	0	0
	9	4	5

We start with the units so that if we have to carry over, we can do so.

7 x 5 = 35, so we put 5 down and carry over the tens (3 in this case).

7 x 3 = 21 Add to that 3 that we carried over and that makes 24. Write 24 down.

Now, we finished multiplying 7 by the top numbers. So will multiply 2 by the top numbers, always starting with units (5 in this case).

The value of the digit 2 is 20. You know that multiplying a whole number by 20, is like doubling and adding a 0 at the end. Rather than doing 20 x 5 and then 20 x 3, we just put a 0 down in the second row of answers. Then 2 x 5 = 10 and 2 x 3 = 6

Adding the two answers: 245 + 700 = 945 So, 35 x 27 = 945

16 Work out:

2 8	3 6	1 9	2 5
x 1 6	x 2 7	x 3 5	x 3 6
+	+	+	+

3 9	3 2	2 8	4 5
x 2 5	x 2 7	x 2 5	x 1 3
+	+	+	+

3 6	2 7	3 6	4 5
x 1 2	x 2 2	x 2 8	x 1 9
+	+	+	+

17

a) Steve is planning a camping trip. He needs to save £75 per month for the trip, for four months. How much will he save in total? ______	b) James has 16 employees. They each work 40 hours a week. How many hours do they work in total each week? ______
c) Helen is investing in a mutual fund. She decides to invest £39 each month. How much will she have invested in 24 months? ______	d) Jenny is paying for a gym membership. The membership costs £32 a month. How much will she pay in total for nine months? ______
e) John is saving for a course. He saves £78 each month. How much will he save in 12 months? ______	f) Asia's phone plan costs £27 a month. Calculate how much it will cost in a year. ______
g) 21 schools are participating in a multicultural festival. Each school sets up five stalls. How many stalls are there in total? ______	h) A computer network has 25 servers. Each server can handle 40 users at any one time. How many users can the network handle in total? ______

Here are some more examples.

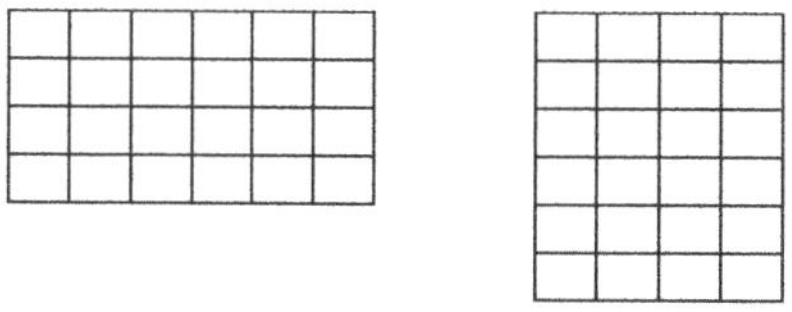

4 x 6 represents the same amount as 6 x 4

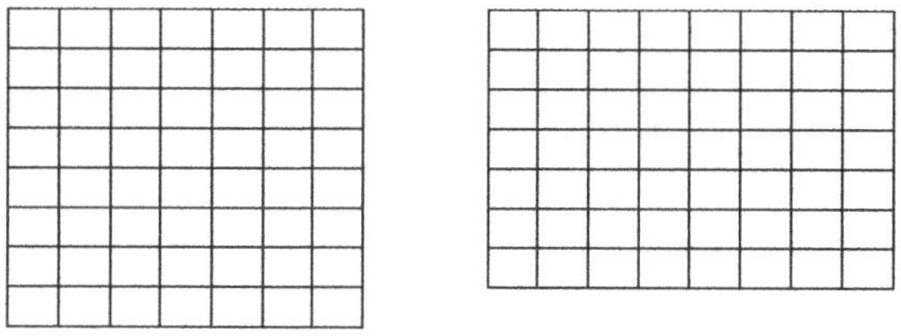

8 x 7 represents the same amount as 7 x 8

There are two ways to work out the total number of chocolate squares.

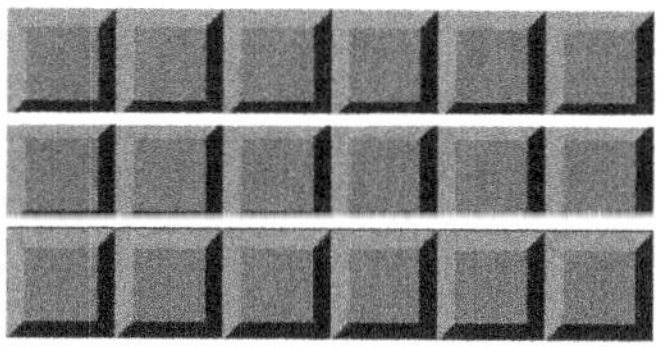

3 rows with 6 squares each:
3 x 6 = 18

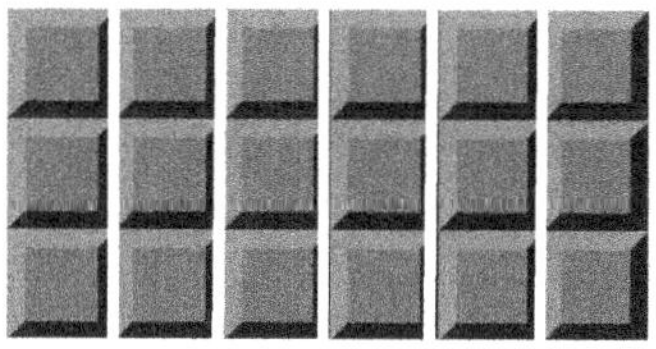

6 columns with 3 squares each:
6 x 3 = 18

18 Fill the gaps.

7 x 5 is the same as 5 x 7	2 x 10 is the same as 10 x ___	12 x 4 is the same as 4 x ___
5 x 6 is the same as 6 x ___	11 x 3 is the same as ___ x 11	7 x 8 is the same as 8 x ___

Order doesn't matter when multiplying.
Example: 4 x 3 = 1 2 3 x 4 = 12

19 Fill the gaps.

3 x ___ = 15	8 x ___ = 40	7 x ___ = 35	___ x 11 = 44	___ x 3 = 21	___ x 5 = 45
6 x ___ = 24	9 x ___ = 54	4 x ___ = 36	___ x 2 = 24	___ x 10 = 60	___ x 7 = 49

20 A digital artist creates 16 designs per week. How many designs will they create in eight weeks?

____

21 A charity donates 21 books to each of the twelve local schools. How many books does the charity donate in total?

____

22 Mark buys a new car and needs car insurance. He finds insurance for £37 a month. How much does this cost in total for 12 months?

____

23 Dea says, '4 x 8 is the same as 2 x 8 and 2 x 82'. Is this correct?

____

24 Heather says, 'Any number multiplied by 0, equals 0'. Is she correct?

____

Progress check

1 Ann works in a shop. She had 26 packs of beans delivered. In each pack, there are 4cans of baked beans.

How many cans are there altogether?

____

2 Maya spends £68 on groceries every week.
How much does Maya spend on groceries in one year? _____ x _____ = _____

3 Fill in the gaps below:

9 x 36 = _____	53 x 10 = _____	6 x _____ = 72	8 x _____ = 64

4 Naila writes these maths sentences:

16 x 4 = 4 x 16	1 x 25 = 1	35 x 0 = 35

Do you agree with Naila? Explain your answer.

____

6 Find another two numbers you can multiply which give the same total as **18 x 36** _____ x _____

Divide three-digit whole numbers by single and double-digit whole numbers and express remainders

We use division when:

- sharing items equally with others
- working out pay per hour, knowing pay per day
- calculating price per item when buying a pack
- sharing food in portions
- working out how long it takes to do one task
- planning exercise routines
- working out the cost of bills per month

Example: Sonia has 14 apples. She shares them equally with her friend, Dan.

14 ÷ 2 = 7 They get 7 apples each.

Example: Ben has £15. He will share it equally with his two friends, Desi and Norma. How much will each of them get?

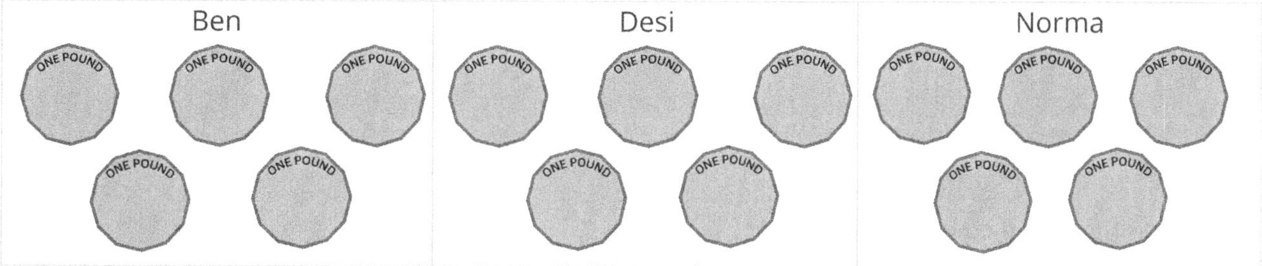

15 ÷ 3 = 5 They get £5 each.

Here is the times tables grid. How can the grid help? What is the link with times tables?

x	0	1	2	3	4	5	6	7	8	9	10	11	12
0	0	0	0	0	0	0	0	0	0	0	0	0	0
1	0	1	2	3	4	5	6	7	8	9	10	11	12
2	0	2	4	6	8	10	12	14	16	18	20	22	24
3	0	3	6	9	12	15	18	21	24	27	30	33	36
4	0	4	8	12	16	20	24	28	32	36	40	44	48
5	0	5	10	15	20	25	30	35	40	45	50	55	60
6	0	6	12	18	24	30	36	42	48	54	60	66	72
7	0	7	14	21	28	35	42	49	56	63	70	77	84
8	0	8	16	24	32	40	48	56	64	72	80	88	96
9	0	9	18	27	36	45	54	63	72	81	90	99	108
10	0	10	20	30	40	50	60	70	80	90	100	110	120
11	0	11	22	33	44	55	66	77	88	99	110	121	132
12	0	12	24	36	48	60	72	84	96	108	120	131	144

48 ÷ 4 = 12
48 ÷ 12 = 4

What do you notice?

Division is the opposite of multiplication.

1 Work out.

10 ÷ 2 = ___	40 ÷ 4 = ___	24 ÷ 6 = ___	42 ÷ 6 = ___	63 ÷ 7 = ___
15 ÷ 5 = ___	32 ÷ 8 = ___	16 ÷ 4 = ___	64 ÷ 8 = ___	100 ÷ 10 = ___
21 ÷ 3 = ___	16 ÷ 2 = ___	18 ÷ 3 = ___	81 ÷ 9 = ___	40 ÷ 5 = ___

A chocolate bar has 18 squares. There are different ways to divide the total number of chocolate squares.

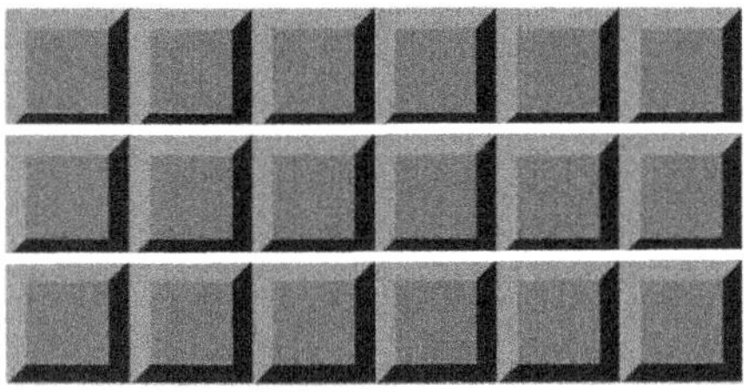

18 ÷ 3 = 6 There are 6 squares in each row.

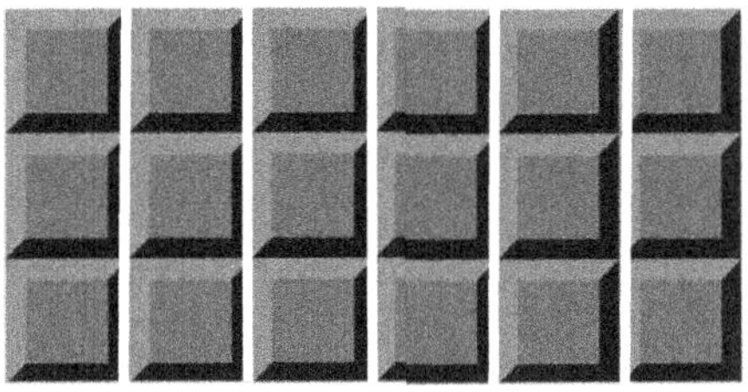

18 ÷ 6 = 3 There are 3 squares in each column.

When dividing large numbers, we start with the tens, then move to the ones.

Example: 42p ÷ 2

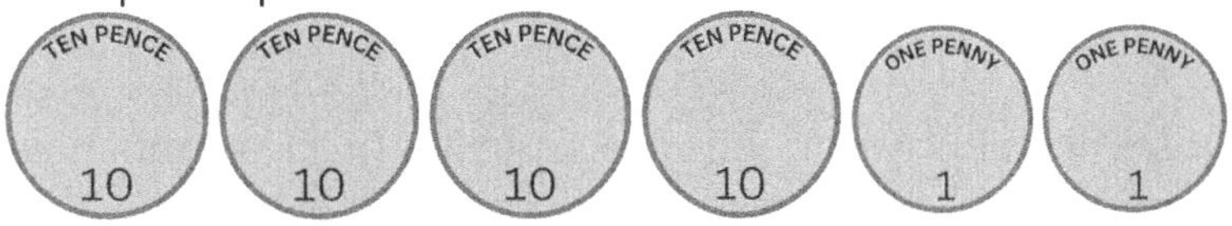

We start by sharing the 10p coins first:

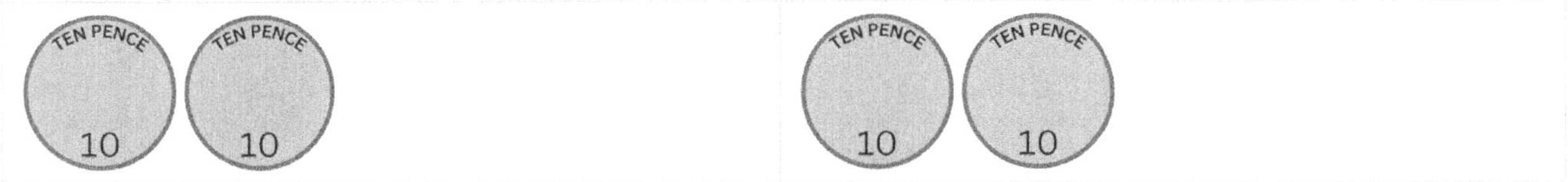

Then sharing the 1p coins too:

Each person gets 21p.

Example: 69 ÷ 3

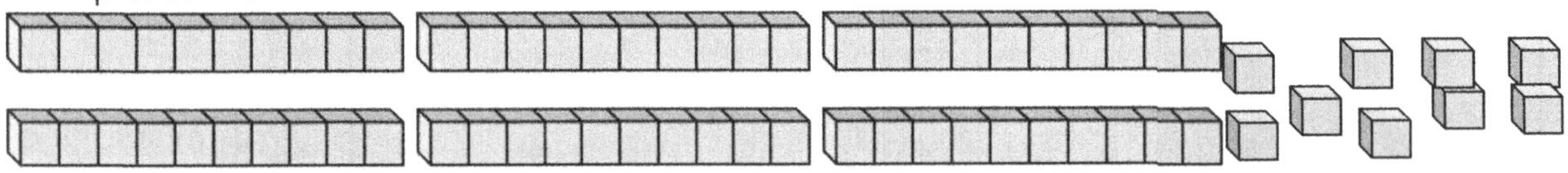

We start by sharing the blocks of 10 first:

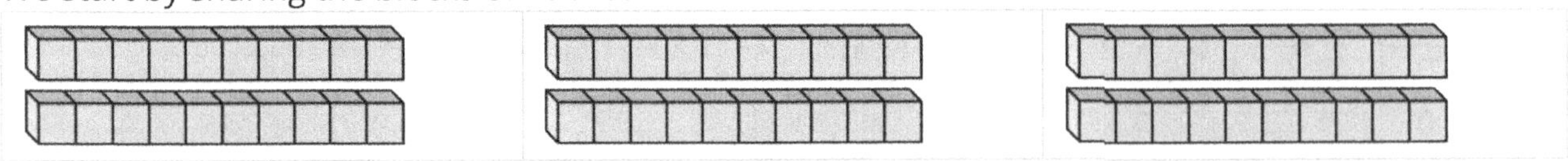

Then sharing the ones too:

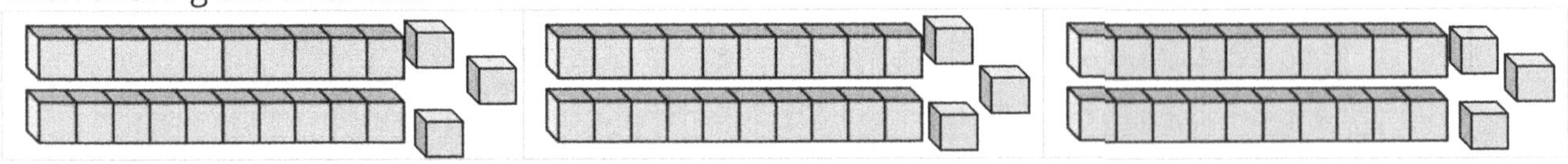

Each person gets 23.

The short division method for division of three-digit by one-digit numbers, without remainders

Example 1

448 ÷ 4

Step 1

Write the calculation correctly.

$$4\overline{\smash{)}\,4\quad 4\quad 8}$$

Visual representation:

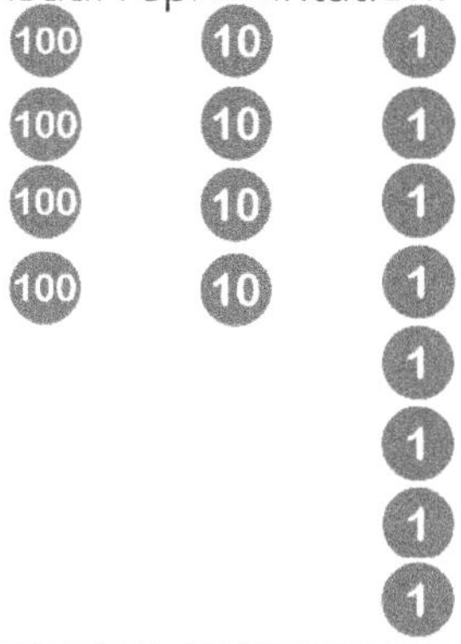

Step 2

How many groups of 4 can we make with the hundreds? Only 1, as 4 ÷ 4 = 1
We write the answer at the top.

$$\begin{array}{r} 1 \\ 4\overline{\smash{)}\,\not{4}\quad 4\quad 8} \end{array}$$

Visual representation:

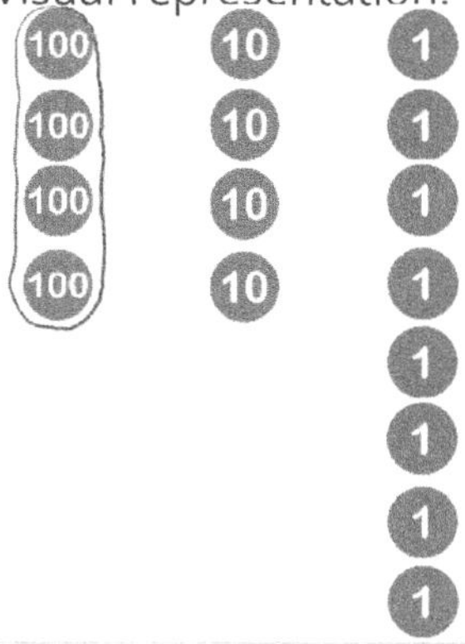

Step 3

How many groups of 4 can we make with the tens? Only 1, as 4 ÷ 4 = 1
We write the answer at the top.

$$\begin{array}{r} 1\quad 1 \\ 4\overline{\smash{)}\,\not{4}\quad \not{4}\quad 8} \end{array}$$

Visual representation:

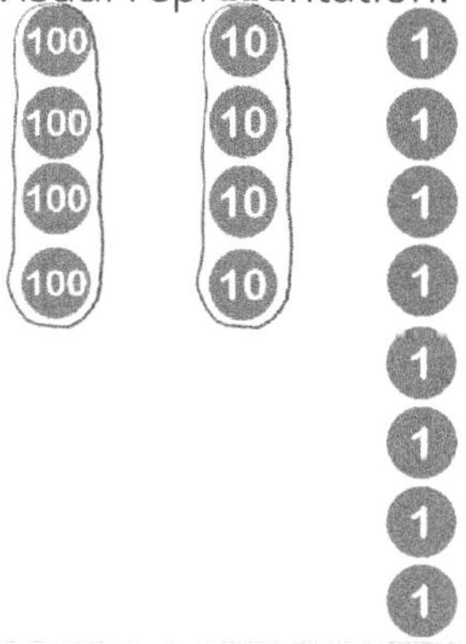

Step 4

How many groups of 4 can we make with the ones? Only 2, as 8 ÷ 4 = 2

We write the answer at the top.

$$\begin{array}{r} 1\quad 1\quad 2 \\ 4\overline{\smash{)}\,\not{4}\quad \not{4}\quad \not{8}} \end{array}$$

There are no more digits left, so 448 ÷ 4 = 112
or 4 x 112 = 448

Visual representation:

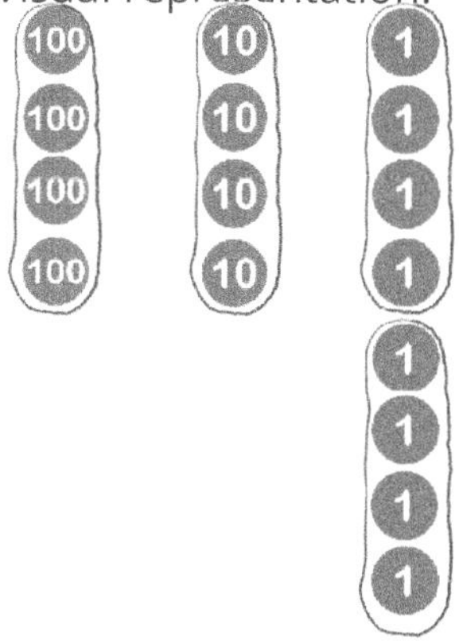

615 ÷ 5

Example 2

Step 1

Write the calculation correctly.

```
5 | 6  1  5
```

Visual representation:

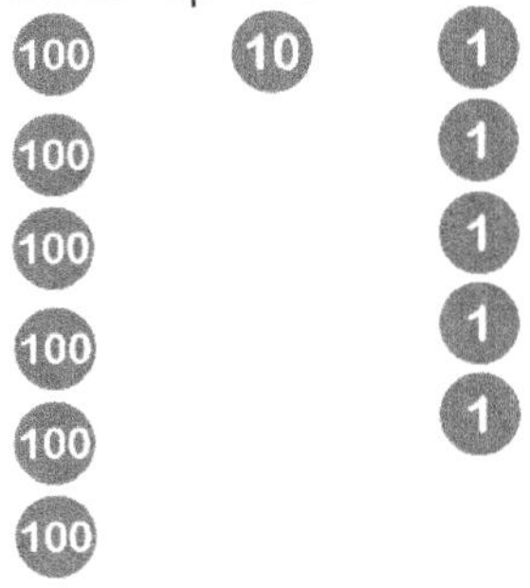

Step 2

How many groups of 5 can we make with the hundreds? Only 1, as 6 ÷ 5 = 1 rem 1.
We write the answer at the top.

```
    1
5 | 6  1  5
```

Visual representation:

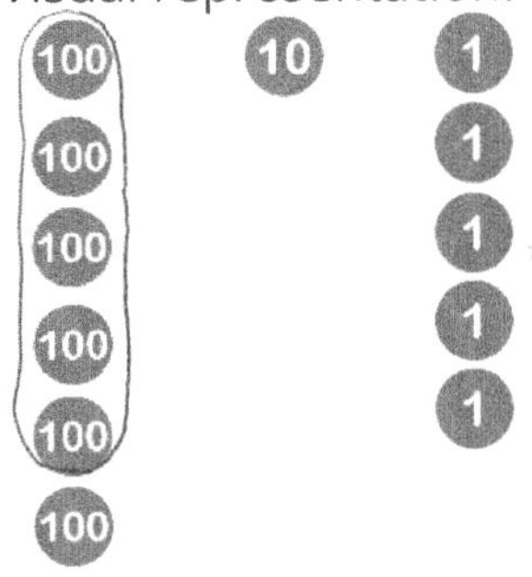

Step 3

How many groups of 5 can we make with the tens? 2, as 11 ÷ 5 = 2 rem 1
We write the answer at the top.

```
    1   2
5 | 6 [1]1 [1]5
```

Visual representation:

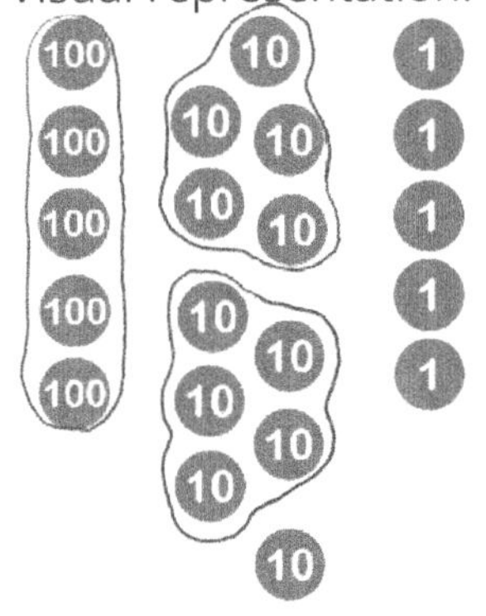

Step 4

How many groups of 5 can we make with the ones? Only 3, as 15 ÷ 5 = 3

We write the answer at the top.

```
    1   2   3
5 | 6 [1]1 [1]5
```

There are no more digits left, so 615 ÷ 5 = 123
or 5 x 123 = 615

Visual representation:

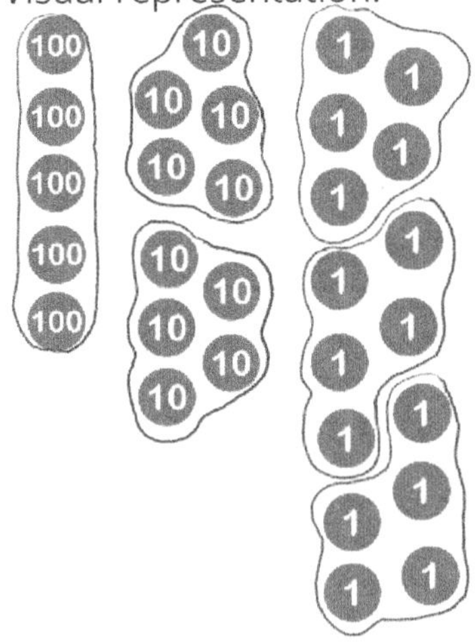

732 ÷ 6

Example 3

Step 1

Write the calculation correctly.

6 | 7 3 2

Visual representation:

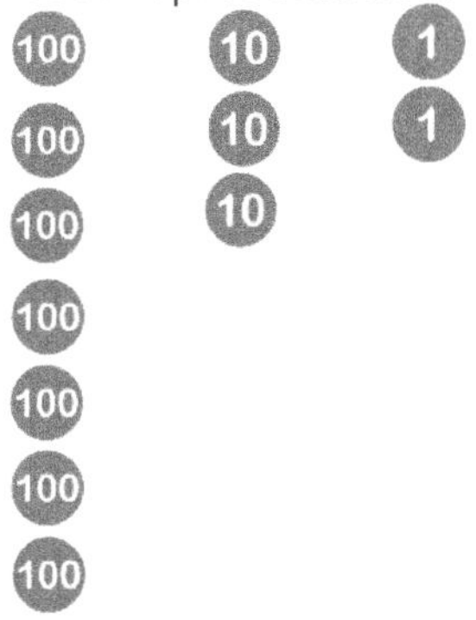

Step 2

How many groups of 6 can we make with the hundreds? Only 1, as 7 ÷ 6 = 1 rem 1

We write the answer at the top.

1

6 | 7 [1]3 2

Visual representation:

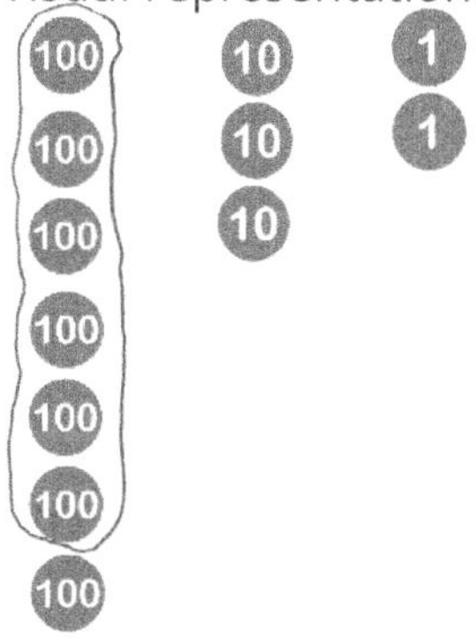

Step 3

How many groups of 6 can we make with the tens? 2, as 13 ÷ 6 = 2 rem 1

We write the answer at the top.

1 2

6 | 7 [1]3 [1]2

Visual representation:

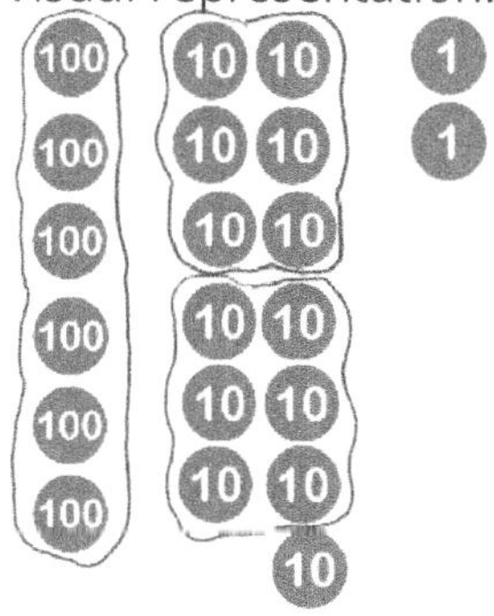

Step 4

How many groups of 6 can we make with the ones? 2, as 12 ÷ 6 = 2

We write the answer at the top.

1 2 2

6 | 7 [1]3 [1]2

There are no more digits left, so 732 ÷ 6 = 122 or 6 x 122 = 732

Visual representation:

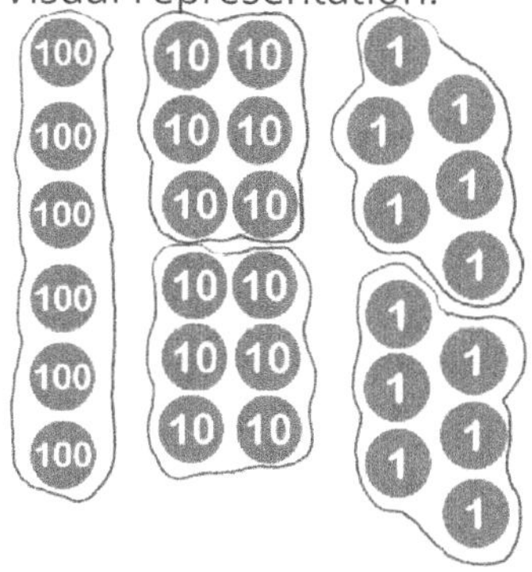

2 Work out. Use a calculator to check your answer.

566 ÷ 2 =	792 ÷ 3 =	956 ÷ 4 =	914 ÷ 2 =	795 ÷ 5 =	840 ÷ 6 =
966 ÷ 7 =	858 ÷ 3 =	680 ÷ 4 =	873 ÷ 3 =	903 ÷ 7 =	856 ÷ 8 =
486 ÷ 3 =	909 ÷ 9 =	675 ÷ 5 =	48 ÷ 3 =	568 ÷ 4 =	444 ÷ 6 =

3

a) Dan spent £360 on groceries over four weeks. He spent the same amount each week. How much did he spend each week? __________	b) Lola has three children. She is putting £498 into a savings account. She will split this amount equally between her children. How much will each child get? __________	c) A group of students raised £165, which they will share equally between five charities. How much will each charity get? __________
d) A football stadium can hold 920 spectators in total. They are hosting four matches simultaneously. The stadium is full and each match is watched by the same number of people. How many people can watch each match? __________	e) Leo works in a bakery. He has 144 cookies. He packs the cookies into boxes. Each box holds eight cookies. How many boxes does Leo need? __________	f) Mini has to distribute 375 charity leaflets. She divides them equally among five volunteers. How many leaflets does each volunteer get? __________

4 Asma is dividing 72 by 3 and thinks the answer is 20. What did she do wrong?

____

Short division method for division of three-digit by one-digit numbers, with remainders

Example 1

745 ÷ 7

Step 1

Write the calculation correctly.

7 | 7 4 5

Step 2

How many groups of 7 can we make with the hundreds? Only 1, as 7 ÷ 7 = 1
We write the answer at the top.

1

7 | ~~7~~ 4 5

Step 3

How many groups of 7 can we make with the tens? 0, as 4 ÷ 7 = 0 rem 4
We write the answer at the top.

1 0

7 | ~~7~~ ~~4~~ [4]5

Step 4

How many groups of 7 can we make with the ones? 6, as 45 ÷ 7 = 6 rem 3

We write the answer at the top.

1 0 6 rem 3

7 | ~~7~~ ~~4~~ [4]~~5~~

There are no more digits left, so 745 ÷ 7 = 106 rem 3

Example 2

657 ÷ 8

Step 1

Write the calculation correctly.

8 | 6 5 7

Step 2

How many groups of 8 can we make with the hundreds? 0, as 6 ÷ 8 = 0 rem 6
We write the answer at the top.

0

8 | ~~6~~ [6]5 7

Step 3

How many groups of 8 can we make with the tens? 8, as 65 ÷ 8 = 8 rem 1
We write the answer at the top.

0 8

8 | ~~6~~ [6]~~5~~ [1]7

Step 4

How many groups of 8 can we make with the ones? Only 2, as 17 ÷ 8 = 2 rem 1

We write the answer at the top.

0 8 2 rem 1

8 | ~~6~~ [6]~~5~~ [1]~~7~~

There are no more digits left, so 657 ÷ 8 = 82 rem 1

5 Work out. Use a calculator to check your answer.

569 ÷ 2 =	798 ÷ 3 =	958 ÷ 4 =	915 ÷ 2 =	798 ÷ 5 =	845 ÷ 6 =
969 ÷ 7 =	860 ÷ 3 =	683 ÷ 4 =	875 ÷ 3 =	908 ÷ 7 =	862 ÷ 8 =
487 ÷ 3 =	914 ÷ 9 =	677 ÷ 5 =	589 ÷ 3 =	567 ÷ 4 =	443 ÷ 6 =

6

a) Paula is preparing packets of pencils for a school supply drive. Each packet contains 7 pencils. She has 215 pencils. How many complete packets can Paula prepare? Show how many pencils are left over. ________________	b) Martin has a book of 257 pages to finish before his English class. He plans to read the same number of whole pages every day for five days. What is the highest number of whole pages he can read each day? How many pages does Martin have left to read? ________________	c) Emily has 794 flyers for her upcoming concert. She gives them to five of her friends to distribute. She divides them equally among her friends. How many flyers does each friend get? How many flyers are left over? ________________
d) Michael has 196 baseball cards. He organises the cards into piles of 9. How many piles does he make? How many cards are left over? ________________	e) Mia works at a charity. She has 350 toys to distribute equally among 15 children's homes. How many toys does each children's home get? How many toys are left over? ________________	f) Karen has 178 apples. She wants to put the apples into crates. She puts 8 apples into each crate. How many crates does Karen need? ________________

7 Asma is dividing 474 by 4 and thinks the answer is 118.2. What did she do wrong?

____

Short division method for division of three-digit by two-digit numbers

Example 1

451 ÷ 11

Step 1

Write the calculation correctly.

11 | 4 5 1

Step 2

How many groups of 11 can we make with the hundreds? None, as 11 does not fit into 4. We write the answer at the top and carry over the 4.

0

11 | ~~4~~ [4]5 1

Step 3

How many groups of 11 can we make with the tens? Only 4, as 45 ÷ 11 = 4 rem 1.
We write the answer at the top.

0 4

11 | ~~4~~ ~~[4]5~~ [1]1

Step 4

How many groups of 11 can we make with the ones? Only 1, as 11 ÷ 11 = 1

We write the answer at the top.

0 4 1

11 | ~~4~~ ~~[4]5~~ ~~[1]1~~

There are no more digits left, so 451 ÷ 11 = 41
or 41 x 11 = 451

Example 2

744 ÷ 12

Step 1

Write the calculation correctly.

12 | 7 4 4

Step 2

How many groups of 12 can we make with the hundreds? None, as 12 does not fit into 7.
We write the answer at the top and carry over the 7.

0

12 | ~~7~~ [7]4 4

Step 3

How many groups of 12 can we make with the tens? 3, as 74 ÷ 12 = 6 rem 2. We write the answer at the top.

0 6

12 | ~~7~~ ~~[7]4~~ [2]4

Step 4

How many groups of 12 can we make with the ones? Only 2, as 24 ÷ 12 = 2

We write the answer at the top.

0 6 2

12 | ~~7~~ ~~[7]4~~ ~~[2]4~~

There are no more digits left, so 744 ÷ 12 = 62
or 12 x 62 = 744

Example 3

958 ÷ 19

Step 1

Write the calculation correctly.

19 | 9 5 8

Step 2

How many groups of 19 can we make with the hundreds? None, as 19 does not fit into 9.
We write the answer at the top and carry over the 9.

0

19 | 9̸ ⁹5 8

Step 3

How many groups of 19 can we make with the tens? 5, as 95 ÷ 19 = 5. We write the answer at the top.

0 5

19 | 9̸ ⁹5̸ 8

Step 4

How many groups of 19 can we make with the ones? None, as 19 does not fit into 8.

We write the answer at the top.

0 5 0 rem 8

19 | 9̸ ⁹5̸ 8̸

There are no more digits left, so:

958 ÷ 19 = 50 rem 8

Example 4

739 ÷ 23

Step 1

Write the calculation correctly.

23 | 7 3 9

Step 2

How many groups of 23 can we make with the hundreds? None, as 23 does not fit into 7.
We write the answer at the top and carry over 7.

0

23 | 7̸ ⁷3 9

Step 3

How many groups of 23 can we make with the tens? 3, as 73 ÷ 23 = 3 rem 4
We write the answer at the top.

0 3

23 | 7̸ ⁷3̸ ⁴9

Step 4

How many groups of 23 can we make with the ones? 2, as 49 ÷ 23 = 2 rem 3

We write the answer at the top.

0 3 2 rem 3

23 | 7̸ ⁷3̸ ⁴9̸

There are no more digits left, so:

739 ÷ 23 = 32 rem 3

When you divide by a two-digit number, round it to estimate the division and then work out the exact number. E.g: round 19 to 20 so you can easily estimate how many 20s fit into 95.

8 Work out. Use a calculator to check your answer.

566 ÷ 12 =	792 ÷ 13 =	956 ÷ 24 =	914 ÷ 22 =	795 ÷ 15 =	840 ÷ 16 =
966 ÷ 27 =	858 ÷ 23 =	680 ÷ 14 =	873 ÷ 13 =	903 ÷ 17 =	856 ÷ 48 =
486 ÷ 33 =	909 ÷ 19 =	675 ÷ 25 =	586 ÷ 53 =	568 ÷ 24 =	444 ÷ 26 =

9

a) John will pay £600 for a 15-month subscription to an online course. How much does the subscription cost per month? ____________	b) Jane bought a laptop for £750. She decides to pay it off over 15 months. How much would she have to pay each month? ____________	c) A company printed 792 brochures for a product launch. They put 24 brochures in each pack. How many packs did they make? ____________
d) Jane needs to finish a novel in two weeks' time. The novel has 240 pages. Jane reads the same number of whole pages every day. What is the highest number of whole pages she can read each day? ____________	e) Leo works in a bakery. He has 150 cupcakes. He puts them into trays of 12. How many full trays will he fill? ____________	f) Mike has 120 apples. He wants to make apple pies. Each apple pie requires 7 apples. How many apple pies can Mike make? How many apples will be left over? ____________

10 Asma is dividing 366 by 4 and thinks the answer is 91 rem 5. What did she do wrong?

____

The long division method

The long division method is like the bus stop method; you just show the working out rather than doing it in your head.

```
      1  6  9
    ---------
5  |  8  4  5
   -  5
      3  4
   -  3  0
         4  5
   -     4  5
            0
```

```
      1  2  7 r 2
    ---------
6  |  7  6  4
   -  6
      1  6
   -  1  2
         4  4
   -     4  2
            2
```

```
      0  9  2 r 3
    ---------
8  |  7  3  9
   -  0
      7  3
   -  7  2
         1  9
   -     1  6
            3
```

```
      0  6  7
    ---------
13 |  8  7  1
   -  0
      8  7
   -  7  8
         9  1
   -     9  1
            0
```

```
      0  2  5
    ---------
24 |  6  0  0
   -  0
      6  0
   -  4  8
      1  2  0
   -  1  2  0
            0
```

```
      0  5  2
    ---------
16 |  8  3  2
   -  0
      8  3
   -  8  0
         3  2
   -     3  2
            0
```

```
      0  3  4 r 3
    ---------
19 |  6  4  9
   -  0
      6  4
   -  5  7
         7  9
   -     7  6
            3
```

```
      0  2  7 r 19
    ---------
25 |  6  9  4
   -  0
      6  9
   -  5  0
      1  9  4
   -  1  7  5
         1  9
```

```
      0  4  9 r 4
    ---------
17 |  8  3  7
   -  0
      8  3
   -  6  8
      1  5  7
   -  1  5  3
            4
```

You can use any division method you like. You don't need to use a specific method. However, it is a good idea to know more than one method.

11

a) Dan uses a pack of butter for five recipes, in equal amounts. A pack of butter weighs 250g. How much butter is Dan using in each recipe? ________	b) The electricity bill for a small business was £810 for six months. How much is the bill per month? ________
c) A school has 435 books to divide equally among 15 classrooms. How many books will each classroom get? ________	d) A bakery made 650 scones to distribute equally among 12 local coffee shops. How many scones did each coffee shop get? ________
e) A farmer has 600 eggs to sell at the local market. He plans to sell them in cartons of 30 eggs. How many cartons can he fill? ________	f) A car hire company has 630 cars spread evenly across 15 branches. How many cars does each branch have? ________
g) A city council spent £882 on planting trees, with each tree costing £18. How many trees were they able to plant? ________	h) A factory produced 756 toys that were distributed equally among 14 stores. How many toys did each store receive? ________

Use a calculator to divide these numbers.

16 ÷ 4 = ____ and 4 ÷ 16 = ____ 10 ÷ 2 = ____ and 2 ÷ 10 = ____

20 ÷ 5 = ____ and 5 ÷ 20 = ____ 25 ÷ 5 = ____ and 5 ÷ 25 = ____

What do you notice?

Order matters when dividing. Example: 4 ÷ 2 = 2 and 2 ÷ 4 = 0.5

12 A pack of eight pens costs £4.80. To work out how much one pen costs in pence. Ola calculates: 480 ÷ 8 = 60p. Is Ola correct?

____

13 A charity runs a food bank and they received a donation of 975 cans of soup. They put 25 cans in each food parcel. To work out how many parcels they will be able to make, Hana works out: 25 ÷ 975 = 39. Is Hana correct?

____

14 A website has 972 pages and the content is equally divided among 18 sections. To work out how many pages there are in each section, Jenny calculates: 972 ÷ 18 = 54. Is Jenny correct?

____

15 Heather says, 'Any number divided by 0, equals 0'. Is she correct?

____

Dividing a number by 0, gives you no answer.
Example: 8 ÷ 0 is undefined. (Imagine having 8 apples and putting them in no groups)

The link between multiplication and division

3 x ___ = 15 is the same as 15 ÷ 3 = ___
___ x 7 = 14 is the same as 14 ÷ 7 = ___

Multiplication ⟷ Division
Multiplication is the opposite of division
Division is the opposite of multiplication

16 Fill the gaps.

6 x ___ = 12 is the same as 12 ÷ 6_____	___ x 7 = 21 is the same as 21 ÷ 7_____
9 x ___ = 27 is the same as _________	___ x 5 = 35 is the same as _________
5 x ___ = 20 is the same as _________	___ x 7 = 42 is the same as _________
8 x ___ = 32 is the same as _________	___ x 11 = 66 is the same as _________

17 Fill the gaps.

3 x 4 = 12 so 12 ÷ 3 = ___ and 12 ÷ 4 = ___	9 x 3 = 27 so 27 ÷ 3 = ___ and 27 ÷ 9 = ___
7 x 5 = 35 so 35 ÷ 5 = ___ and 35 ÷ 7 = ___	10 x 6 = 60 so 60 ÷ 10 = ___ and 60 ÷ 6 = ___
6 x 8 = 48 so 48 ÷ 6 = ___ and 48 ÷ 8 = ___	8 x 11 = 88 so 88 ÷ 8 = ___ and 88 ÷ 11 = ___
4 x 7 = 28 so 28 ÷ 4 = ___ and 28 ÷ 7 = ___	4 x 12 = 48 so 48 ÷ 4 = ___ and 48 ÷ 12 = ___

The link between addition, subtraction, multiplication and division

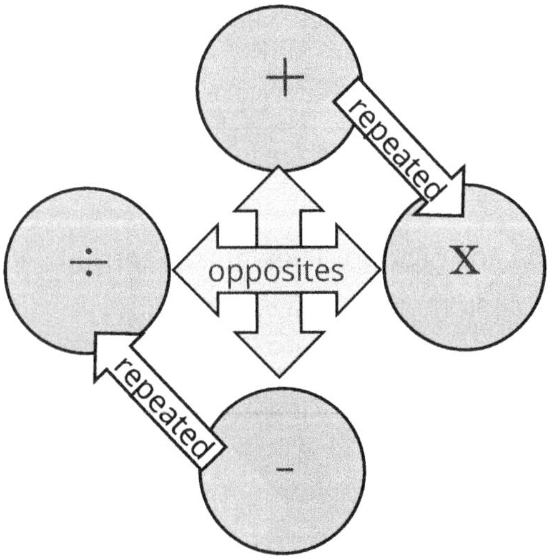

Subtraction is the opposite of **addition**.
Example: 4 + 3 = 7   7 – 3 = 4

Division is the opposite of **multiplication**.
Example: 4 x 3 = 12   12 ÷ 4 = 3

Multiplication is repeated **addition**.
Example: 3 x 4 is the same as 4 + 4 + 4

Division is repeated **subtraction**.
Example: 12 ÷ 3 is the same as 12 - 4 - 4 - 4

We use the opposite calculation usually, to check an answer.

To check addition, we use subtraction. To check subtraction, we use addition. To check multiplication, we use division or repeated addition. To check division, we use multiplication or repeated subtraction.

Checking answers

Example: Danny bought a dishwasher for £369 and a microwave for £199. How much did the items cost in total? 369 + 199 = 568   To check the answer: 568 – 199 = 369 or 568 – 368 = 199

Example: Oliver has £800 in his bank account. He spent £740 on a sofa. How much money does he have in the account after? 800 – 740 = 60   To check: 60 + 740 = 800 or 740 + 60 = 800

Example: An airline operates 780 flights per month. If the flights are evenly distributed over 26 days, how many flights operate each day? 780 ÷ 26 = 30   To check: 30 x 26 =780 or 26 x 30 = 780

Example: Beth bakes 24 muffins. Each muffin needs 15 blueberries. How many blueberries does Beth need in total? 24 x 15 = 360   To check: 360 ÷ 15 = 24 or 360 ÷ 24 = 16

18 Jim moved to a new house. These are the amounts he spends each month on mortgage and utilities. Mortgage: £750 Utilities: £210 Calculate the total of these amounts.

____

Check your answer: _______________

19 Steve is going on a trip. He has a budget of £319. Accommodation costs £76. How much does he have left after paying for accommodation? _______________

Check your answer: _______________

20 Mark works at a bakery. He makes loaves of bread. Each loaf needs 12g of seeds. Mark has 150g of seeds. How many loaves can Mark make with the seeds he has? _______________

Check your answer: _______________

21 Sukhdev buys a new car and needs car insurance. He finds insurance for £39 a month. How much does this cost in total for 12 months? _______________

Check your answer: _______________

22 Sarah has 7 months to save for her college tuition. She saves £120 each month. How much money does she save in 7 months? _______________

Check your answer: _______________

23 Nikki is preparing for an exam. She needs to study a book with 248 pages before the exam. She will study the same number of whole pages every day, for eight days. What is the highest number of whole pages she can read each day? _______________

Check your answer: _______________

You only need to use one method to check your answer.

24 Check these calculations.

18 + 18 = 36 Check: ______	28 - 16 = 12 Check: ______
5 x 4 = 20 Check: ______	29 + 36 = 65 Check: ______
10 + 68 = 78 Check: ______	42 ÷ 7 = 6 Check: ______
56 ÷ 8 = 7 Check: ______	37 - 6 = 31 Check: ______

25 Nisa says: 'For multiplication you have two ways of checking: division and addition'. Is she correct?

______

26 Liam will run 132 miles in 12 weeks. He will run the same distance every week. He works out how many miles he will run per week: 12 ÷ 132 = 11 Is Liam correct?

______

Progress check

1 Ann works in a shop. She had 124 cans of beans delivered. In each pack, there are 4cans of baked beans.

How many packs were delivered?

______

2 Maya spends £43 on groceries every week. How many weeks would £688 last for?

____ ÷ ____ = ____

3 Fill in the gaps below:

360 ÷ 9 = ____	500 ÷ 10 = ____	641 ÷ 7 = ____	637 ÷ ____ = 7

4 Naila writes these maths sentences:

160 ÷ 4 = 4 ÷ 160

500 ÷ 1 = 1

Do you agree with Naila? Explain your answer.

______

5 Jo worked out that **420 ÷ 7 = 60**. How can she check her answer by multiplying?

____ x ____ = ____

Approximate by rounding numbers less than 1000 to the nearest 10 or 100 and use this rounded answer to check results

We round when:

- working out the total cost of items we have bought
- checking that we have done calculations correctly
- estimating the cost of bills over time
- estimating the time it takes to travel to a place

Rounding to the nearest 10

To round to the nearest 10, means to decide which of the tens the number is closest to.

Example: 39

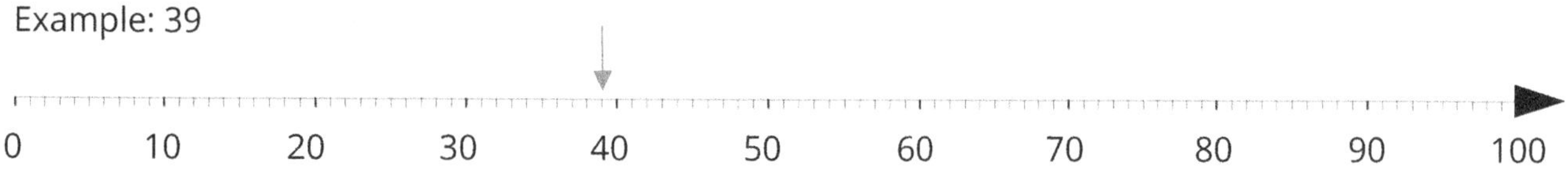

39 is between 30 and 40. As 39 is closer to 40, **it will be rounded to 40**.

Example: 52

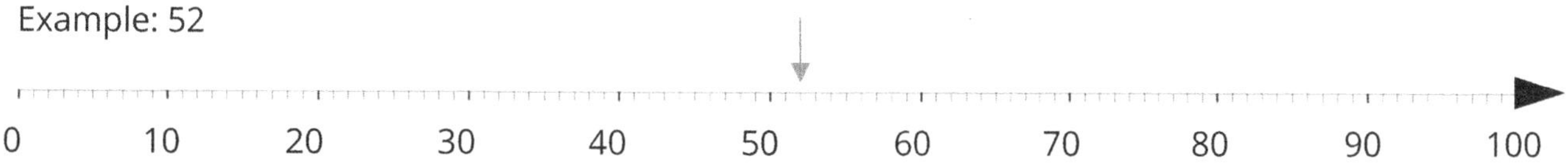

52 is between 50 and 60. As 52 is closer to 50, **it will be rounded to 50**.

Example: 65

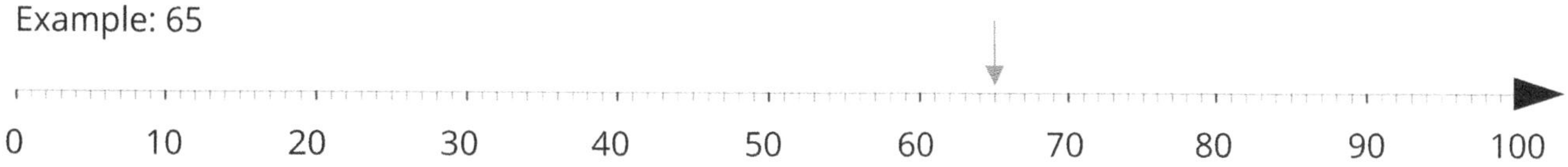

65 is between 60 and 70. **To the nearest ten that is 70**.

Example: 30

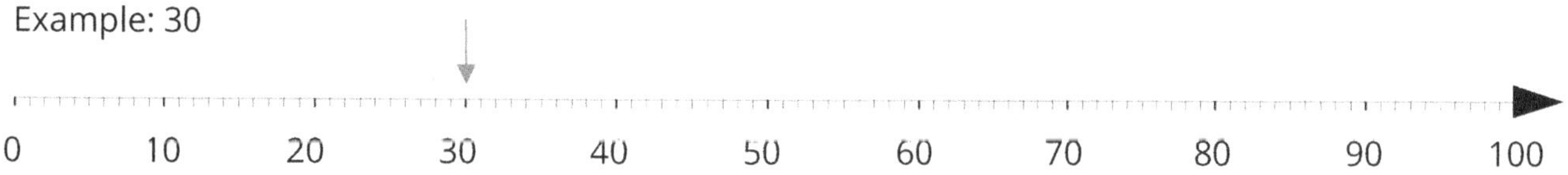

30 is between 30 and 40 or between 20 and 30. **To the nearest ten that is 30**.

Example: 97

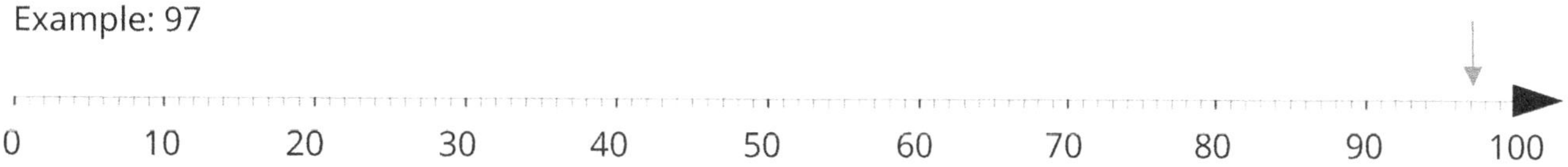

97 is between 90 and 100. As 97 is closer to 100, **it will be rounded to 100**.

Example: 2

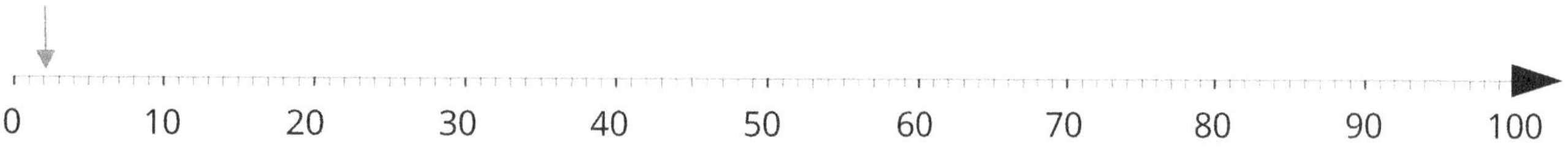

2 is between 0 and 10. As 2 is closer to 0, **it will be rounded to 0**.

To round a number to the nearest ten, you decide by looking at the ones.
If ending in **0**, **1**, **2**, **3** or **4**, you round the number **down**.
If ending in **5**, **6**, 7, **8** or **9**, you round the number **up**.
Example: 42 is rounded to 40. 85 is rounded to 90.

Example: 182

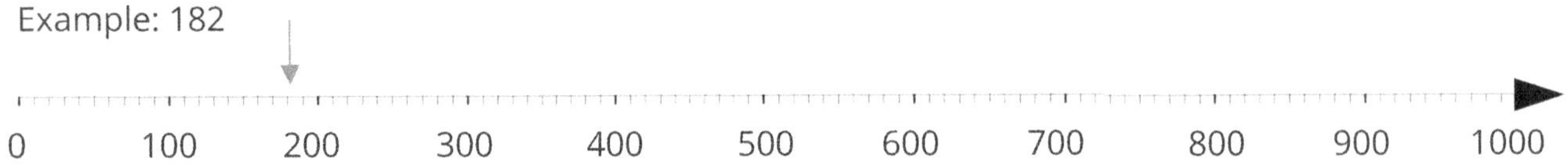

182 is between 180 and 190. As 182 is closer to 180, **it will be rounded to 180**.

Example: 255

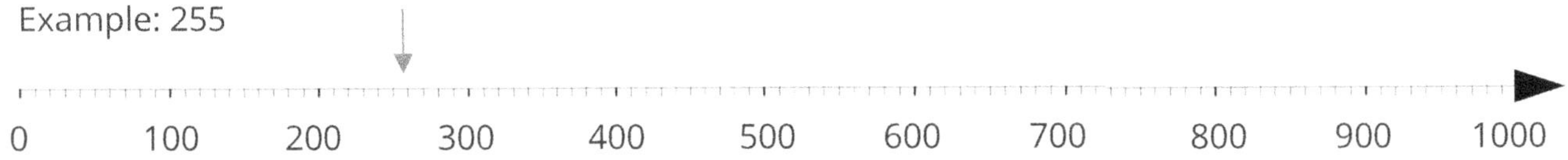

255 is between 250 and 260. As 255 is closer to 260, **it will be rounded to 260**.

Example: 703

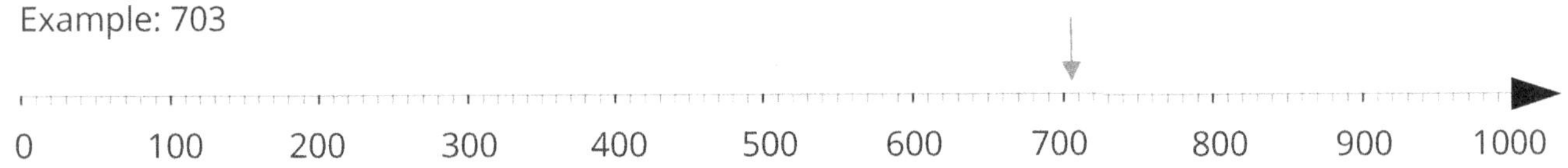

703 is between 700 and 710. As 703 is closer to 700, **it will be rounded to 700**.

Example: 490

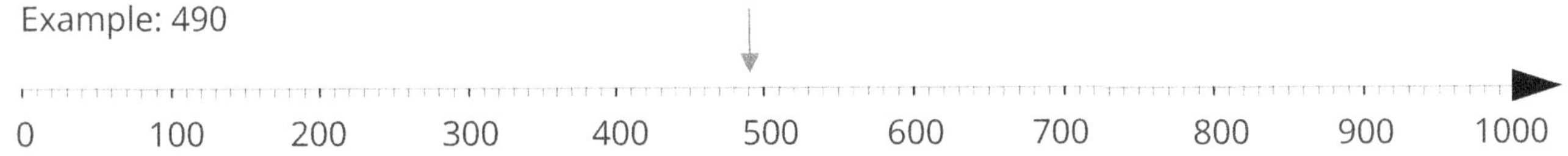

490 is spot on 490. **To the nearest ten that is 490**.

1 Round these numbers to the nearest 10.

83	____	119	____	4	____	427	____	835	____	550	____
276	____	54	____	790	____	599	____	112	____	75	____
963	____	271	____	324	____	68	____	566	____	105	____

2 Tanya is shopping. She bought a bread-maker for £127, a microwave for £143 and a kettle for £45. Use rounding to estimate the total cost of all three items.

___

When estimating, you round all numbers involved to the same degree of accuracy.
E.g: 472 + 86 ➡ 470 + 90 = 560 So both numbers are rounded to the nearest 10.

3 Lana bought a laptop for £379.00. She had a budget of £500. Estimate how much money she has left.

___

4 Afzal works 17 hours a week. Estimate the number of hours he does in a year, if he works every week of the year.

____

5 Luna needs 43 cupcakes for a party. There are 12 cupcakes in one box. Estimate how many boxes she needs to buy.

____

6 Check these calculations using estimation. The first one has been done for you. You can use a calculator.

43 + 38 = Checking: 40 + 40 = 80	646 – 182 = Checking: ______________	116 + 578 = Checking: ______________	933 – 37 = Checking: ______________
647 + 285 = Checking: ______________	559 – 442 = Checking: ______________	170 + 189 + 535 = Checking: ______________	259 - 75 = Checking: ______________

7 Betty has £35. She wants to buy as many candles as possible. One candle costs £8. Use estimation to work out how many candles Betty can buy.

____

8 In a large pack there are 12 loo rolls. To work out how many loo rolls there are in 9 packs, Zen is doing this calculation: 12 x 9 = 108 Use estimation to check this answer.

____

9 A number has been rounded to the nearest ten. The rounded number is 60. What could the original number have been, before it was rounded?

_____, _____, _____, _____, _____, _____, _____, _____, _____, _____

10 A number has been rounded to the nearest ten. The rounded number is 180. What could the original number have been, before it was rounded?

_____, _____, _____, _____, _____, _____, _____, _____, _____, _____

11 Tom's electricity bill is £148, the water bill is £87 and his gas bill is £162 for the month. Estimate the total monthly bill for electricity, water and gas.

____

Rounding to the nearest 100

To round to the nearest 100, means to decide which of the hundreds the number is closest to.

Example: 69

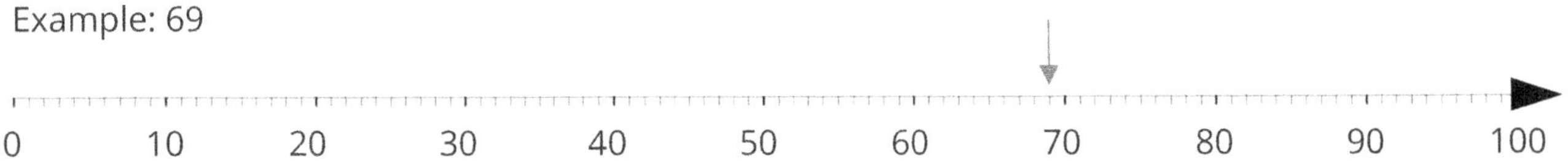

69 is between 0 and 100. As 69 is closer to 100 than to 0, **to the nearest hundred that is 100**.

> To round a number to the nearest hundred, you decide by looking at the tens.
> If ending in **0**, **1**, **2**, **3** or **4**, you round the number **down**.
> If ending in **5**, **6**, 7, **8** or **9**, you round the number **up**.
> Example: 426 is rounded to 400. 850 is rounded to 900.

Example: 42

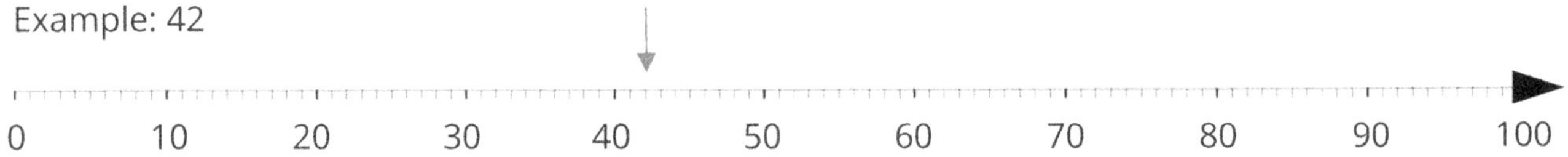

42 is between 0 and 100. As 42 is closer to 0, **it will be rounded to 0**.

Example: 50

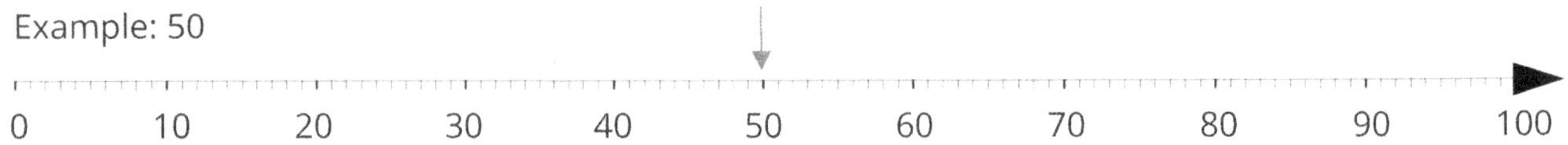

50 is between 0 and 100. As 50 is the cut off point for rounding up, **it will be rounded to 100**.

Example: 100

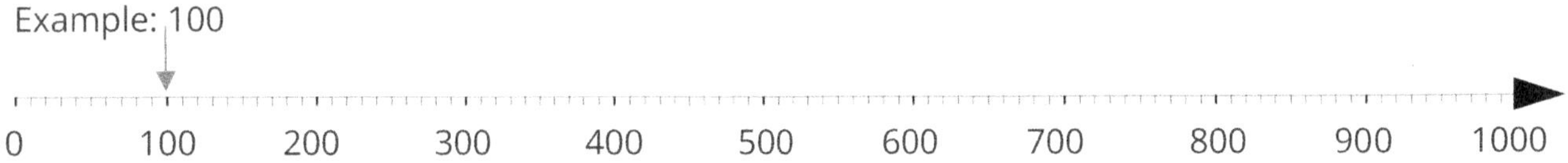

100 is spot on 100. **To the nearest hundred that is 100**.

Example: 197

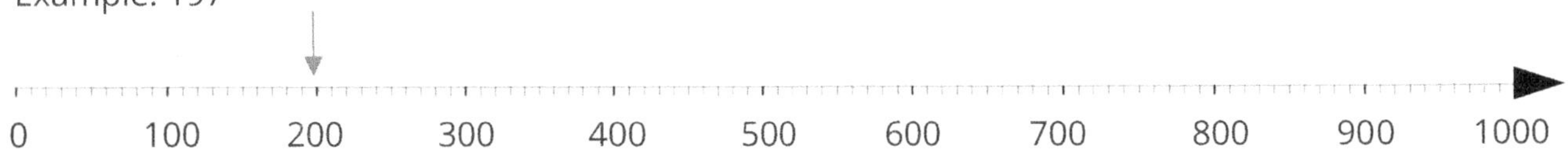

197 is between 100 and 200. As 197 is closer to 200 than to 100, **it will be rounded to 200**.

Example: 249

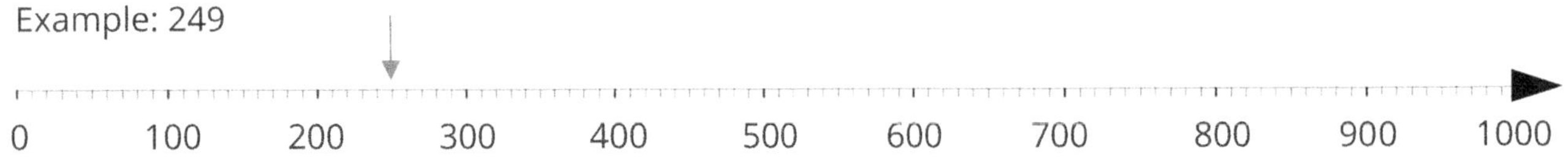

249 is between 200 and 300. As 249 is closer to 200 than to 300, **it will be rounded to 200**.

Example: 310

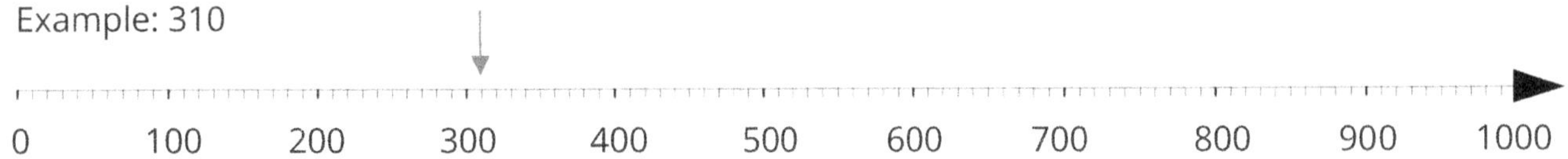

310 is between 300 and 400. As 310 is closer to 300 than to 400, **it will be rounded to 300**.

Example: 650

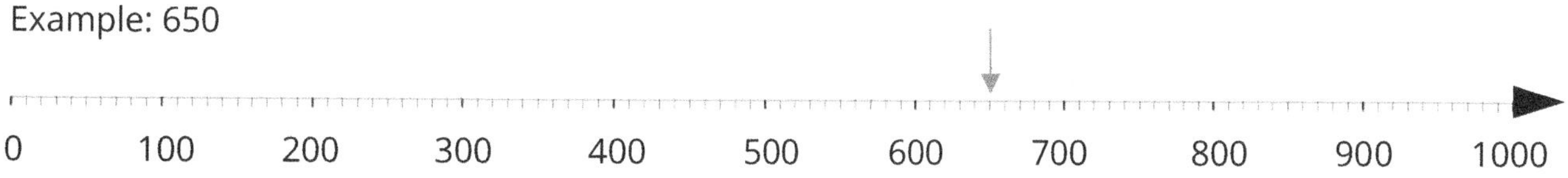

650 is between 600 and 700. As 650 is the cut off point for rounding up, **it will be rounded to 700**.

Example: 800

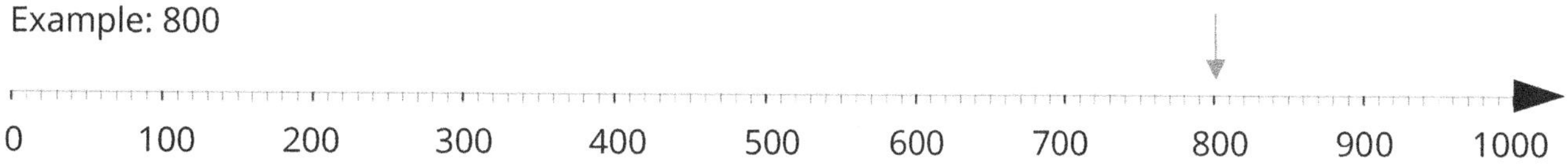

800 is spot on 800. **To the nearest ten that is 800**.

Example: 967

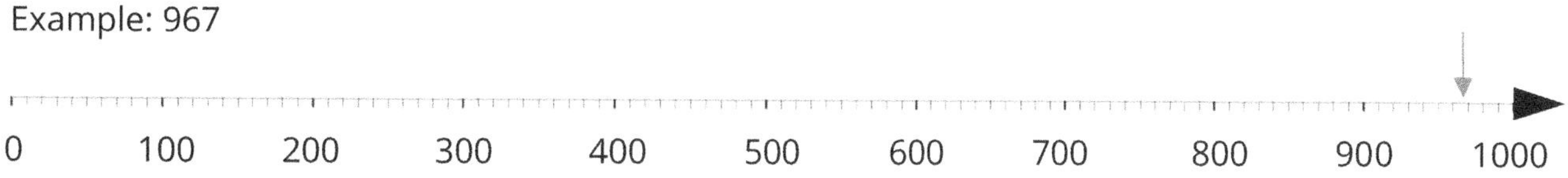

967 is between 900 and 1000. As 967 is closer to 1000 than to 900, **it will be rounded to 1000**.

12 Round these numbers to the nearest 100.

83 _____	119 _____	4 _____	427 _____	835 _____	550 _____
276 _____	54 _____	790 _____	599 _____	112 _____	75 _____
963 _____	271 _____	324 _____	68 _____	566 _____	105 _____

13 Check these calculations using estimation. The first one has been done for you. You can use a calculator.

143 + 398 = Checking: 100 + 400 = 500	674 – 198 = Checking: _______________	131 + 557 = Checking: _______________	973 – 13 = Checking: _______________
640 + 255 = Checking: _______________	554 – 482 = Checking: _______________	140 + 189 + 545 = Checking: _______________	705 – 129 = Checking: _______________

14 A number has been rounded to the nearest hundred. The rounded number is 300. What could the original number have been, before it was rounded?

Highest possible number: _____ Lowest possible number: _____

15 Tom's electricity bill is £148, and his gas bill is £162 for the month. Estimate the total monthly expenses for electricity and gas to the nearest hundred pounds.

____

16 In a charity event, two participants raised £259 and £474 each.

a) Calculate the total amount raised.

____

b) Use estimation to check your answer to part a.

____

17 A mobile phone plan costs £18 per month, and Sam wants to pay for a full year in advance.

a) Calculate the annual cost.

____

b) Round the monthly cost to the nearest 10.

____

18 Round the following numbers to the nearest 10 and then add them: 132, 157, 186. What is the estimated total?

____

19 If you have distances of 34 km, 78 km, and 59 km, round each to the nearest 10 km and find the total estimated distance.

____

20 Round 742, 658, and 395 to the nearest 100. Subtract the smallest from the largest rounded number.

____

21 A school has 312 students, and each student needs a notebook costing £12. Round the numbers to the nearest 10 and estimate the total cost.

____

22 You have a budget of £500 for a party. Estimate the costs of decorations (£123), food (£245), and music (£87) by rounding to the nearest 10. Will you stay within budget?

____

23 Estimate your monthly expenses by rounding to the nearest 100: Rent - £650, Groceries - £230, Utilities - £120.

____

24 Estimate the sum of 823 and 275 by rounding each number to the nearest 100.

___

25 You're planning a road trip. The first leg is 265 miles, and the second leg is 378 miles. Round each leg to the nearest 100 miles and estimate the total distance.

___

26 To check her subtraction of 637 - 258, Emma rounds each number to the nearest 100 (600 and 200) and calculates 600 - 200 = 400. She concludes her answer of 379 is incorrect. What mistake did Emma make in her rounding process?

___

Progress check

1 Fatima received a salary of £296 the 1st week and £298 the 2nd week. How much did she get in total? Check your answer using estimation.

___

2 Jade spent £281 the 1st week, £319 the 2nd week, and £356 the 3rd week. How much did she spend in total? Check your answer using estimation.

___

3 A number has been rounded to the nearest ten. The rounded number is 60. What could the original number have been, before it was rounded?

____, ____, ____, ____, ____, ____, ____, ____, ____, ____

4 A number has been rounded to the nearest hundred. The rounded number is 500. What could the original number have been, before it was rounded?

Highest possible number: ____ Lowest possible number: ____

5 Jade rounded 30 to the nearest 10. Here is her answer: 30. Is Jade correct? Explain your answer.

___

6 a) Round 43 to the nearest 10. ___

b) Round 43 to the nearest 100. ___

c) Round 430 to the nearest 10. ___

d) Round 430 to the nearest 100. ___

Read, write and understand thirds, quarters, fifths and tenths including equivalent forms

We use fractions when:

- sharing food
- cooking
- splitting the bill
- working with time
- doing DIY activities

$$\frac{1}{4}$$

1 — how many parts you are taking

4 — how many parts there are in total

A fraction is part of a whole.

Whole	Half	Third	Quarter	Fifth	Tenth
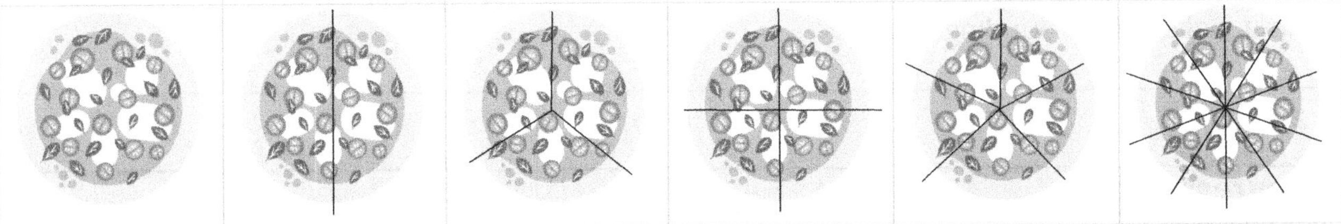					
1 part	2 parts	3 parts	4 parts	5 parts	10 parts

A half is written as $\frac{1}{2}$ and means 1 out of 2. It can also be shown as: 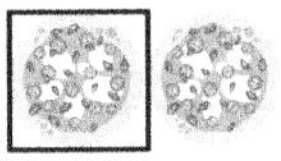

A third is written as $\frac{1}{3}$ and means 1 out of 3. It can also be shown as: 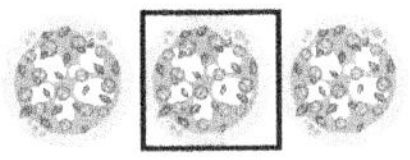

A quarter is written as $\frac{1}{4}$ and means 1 out of 4. It can also be shown as: 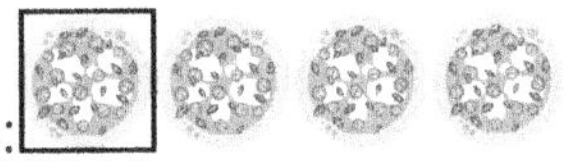

A fifth is written as $\frac{1}{5}$ and means 1 out of 5. It can also be shown as: 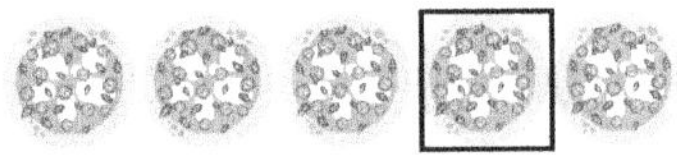

A tenth is written as $\frac{1}{10}$ and means 1 out of 10. It can also be shown as: 

If we cut a pie into 3 equal parts and take 1 part, we've taken a third of it.

The pie can be cut in different ways, as long as it is split into three **equal parts**.

These are also examples of shapes cut into thirds:

1 Cut these shapes into thirds.

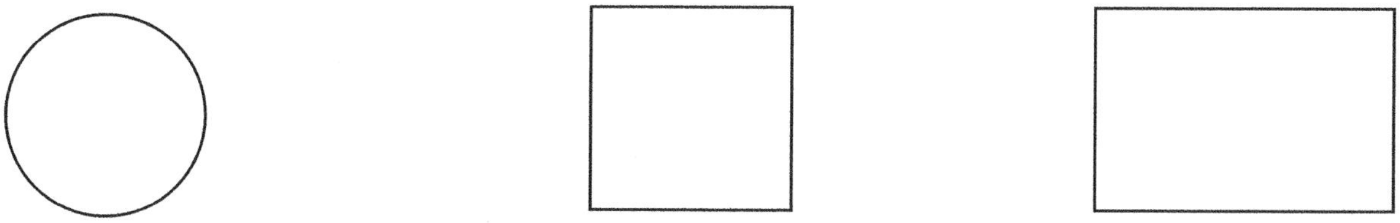

2 Lynn thinks she has cut this shape into thirds.
Is she correct? Why?

3 Shade $\frac{1}{3}$ of this shape.

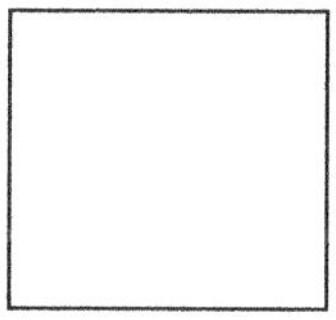

One third	Two thirds	Three thirds
$\frac{1}{3}$	$\frac{2}{3}$	$\frac{3}{3}$
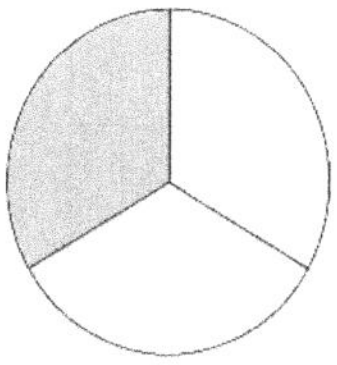	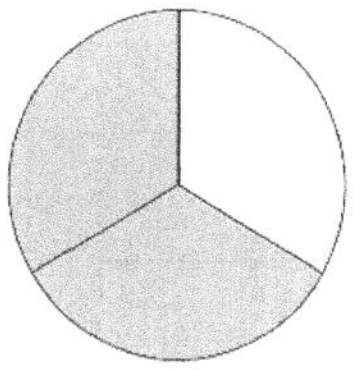	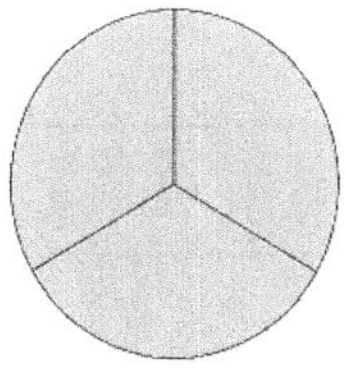
1 out of 3	2 out of 3	3 out of 3

4 Shade $\frac{2}{3}$ of this shape.

5 Compare these fractions using >, < or = .

$\frac{1}{3} \bigcirc \frac{2}{3}$	$\frac{3}{3} \bigcirc \frac{2}{3}$	$\frac{3}{3} \bigcirc \frac{1}{3}$

6 Tick one third of these cupcakes.

To work out a third, you divide the number by 3.

7 a) Desi will use a third of these eggs to make a cake.

How many eggs will Desi use?

____

b) Nike has £30.

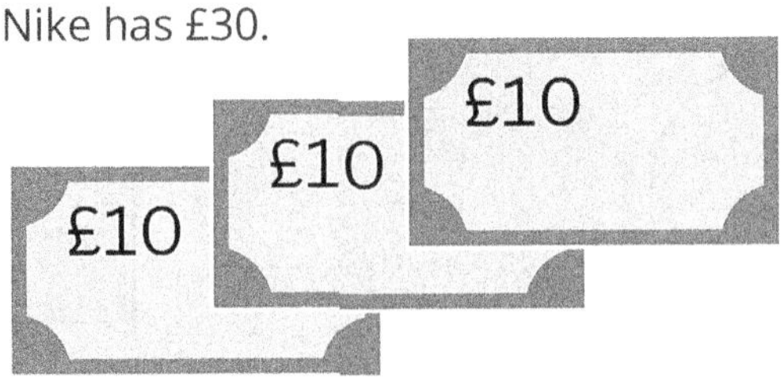

She spent a third of this amount on a new notebook. How much did the notebook cost?

____

8 What fraction of each shape is coloured in?

$\frac{2}{6}$ or $\frac{\quad}{3}$

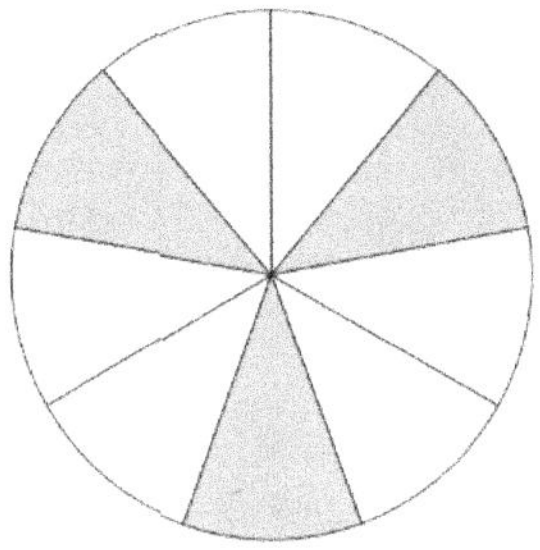

$\frac{\quad}{9}$ or $\frac{1}{3}$

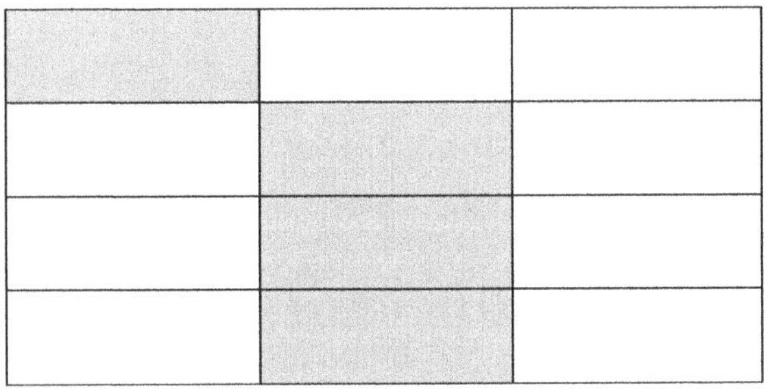

$\frac{\quad}{12}$ or $\frac{\quad}{3}$

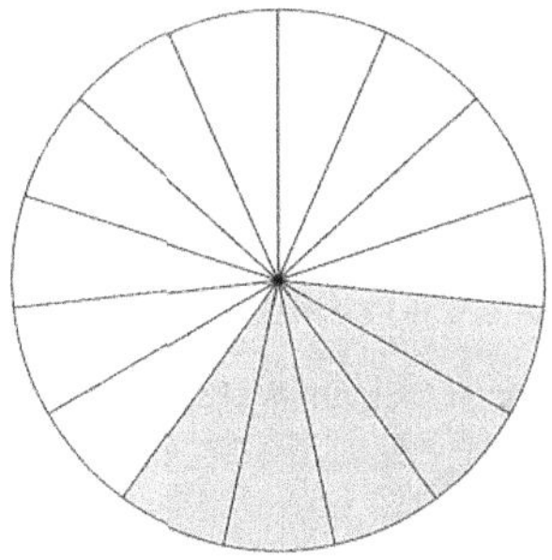

$\frac{\quad}{15}$ or $\frac{\quad}{3}$

If we cut a pie into 4 equal parts and take 1 part, we've taken a quarter of it.

The pie can be cut in different ways, as long as it is split into four equal parts.

These are also examples of shapes cut into quarters:

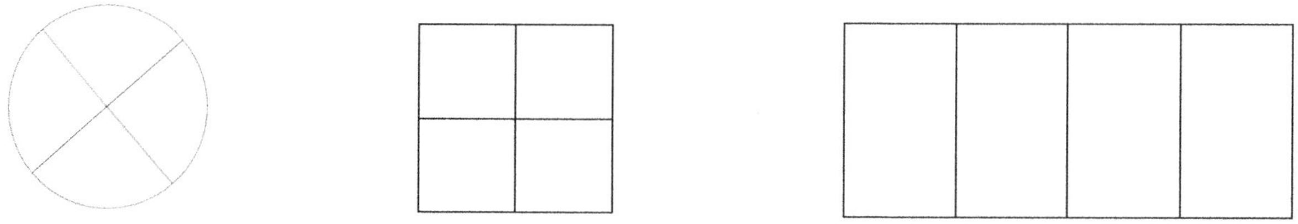

9 Cut these shapes into quarters.

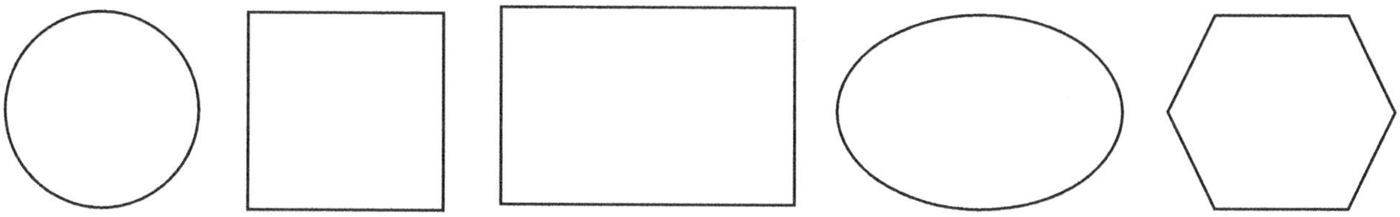

10 Lynn thinks she has cut this shape into quarters.

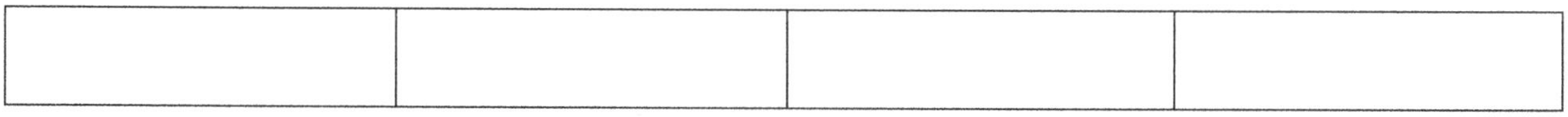

Is she correct? Why?

____

11 Shade $\frac{1}{4}$ of this shape.

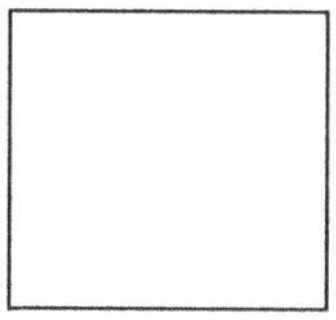

12 Shade $\frac{2}{4}$ of this shape.

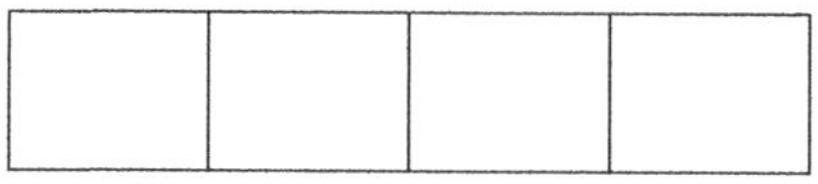

One quarter	Two quarter	Three quarters	Four quarters
$\frac{1}{4}$	$\frac{2}{4}$	$\frac{3}{4}$	$\frac{4}{4}$
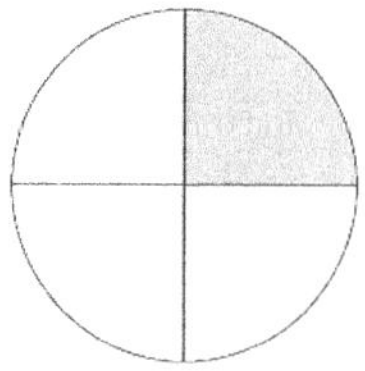	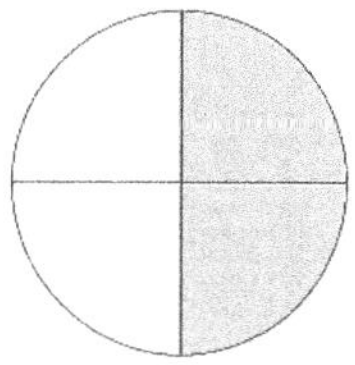	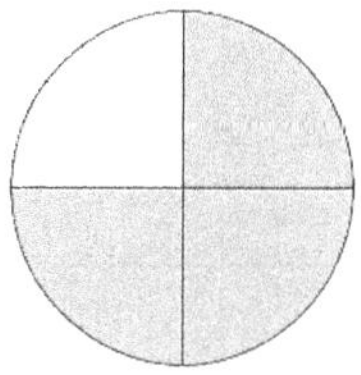	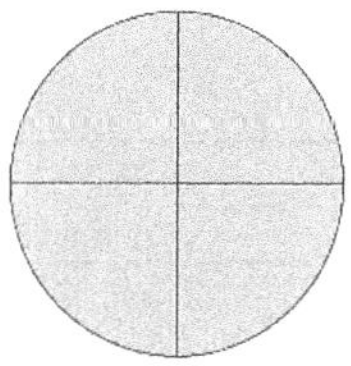
1 out of 4	2 out of 4	3 out of 4	4 out of 4

13 Compare these fractions using >, < or =

$\frac{2}{4} \bigcirc \frac{3}{4}$	$\frac{1}{4} \bigcirc \frac{4}{4}$	$\frac{2}{4} \bigcirc \frac{1}{4}$

To work out a quarter, you divide the number by 4.

14 a) Sue is sharing this pizza into quarters.

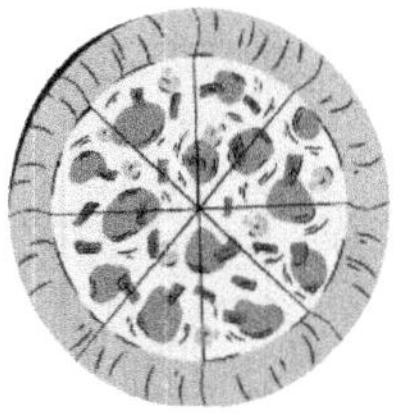

How many slices will there be in each quarter?

____

b) Steph has 12 sweets.

She is giving a quarter of them to her friend. How many sweets is Steph giving to her friend?

____

c) Jo is donating a quarter of these cans of tuna to a foodbank.

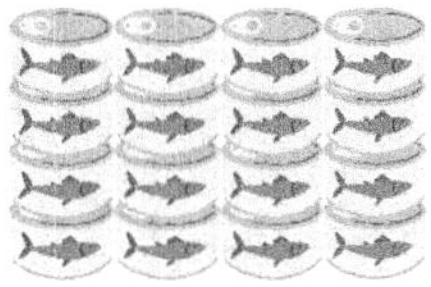

How many cans of tuna is Jo donating?

____

d) Nike has £20.

She will give a quarter of the money to her friend. How much will she give to her friend?

____

e) Sue is sharing this cake into quarters. How many cherries will be in each quarter?

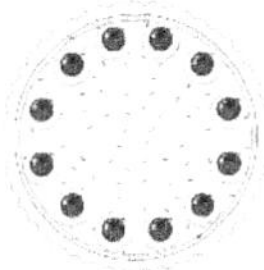

____

15 What fraction of the shapes is coloured in?

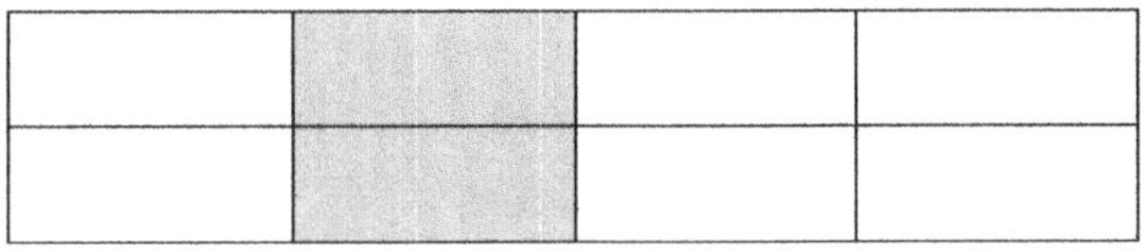

$\frac{\ }{8}$ or $\frac{\ }{4}$

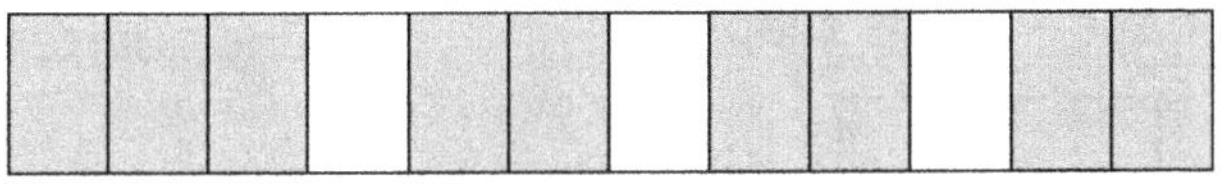

$\frac{\ }{12}$ or $\frac{\ }{4}$

$\frac{\ }{16}$ or $\frac{\ }{4}$

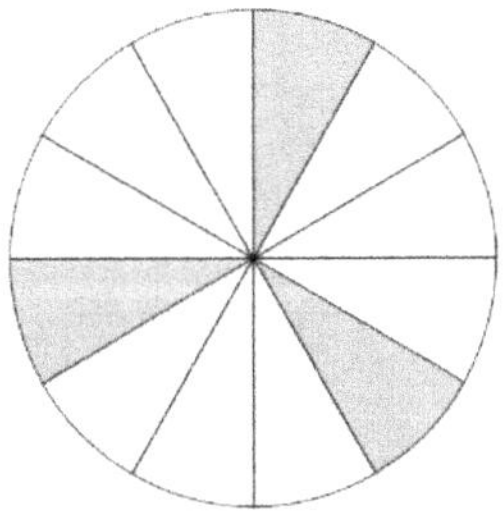

$\frac{\ }{12}$ or $\frac{\ }{4}$

If we cut this pie into **5 equal parts** and take 1 part, we've taken **a fifth** of it.

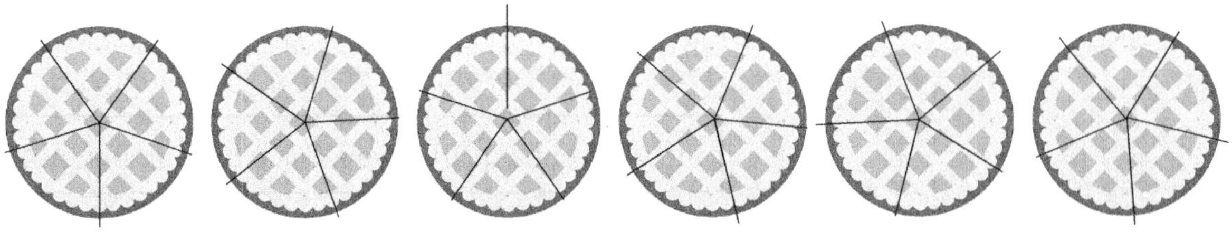

The pie can be cut in different ways, as long as it is split into five equal parts.

These shapes have been cut into fifths.

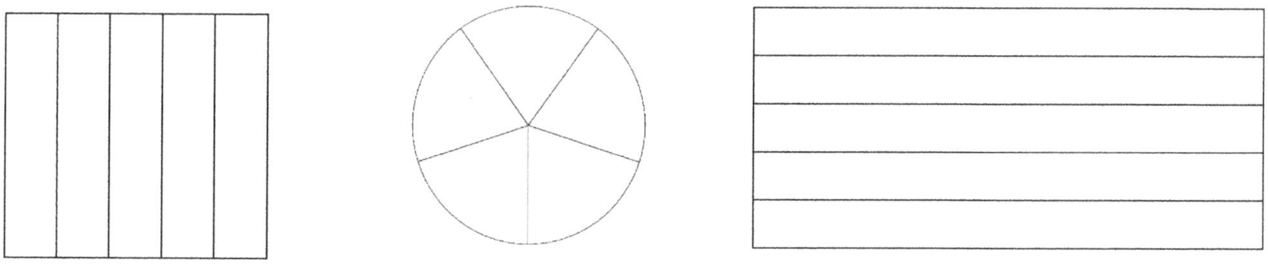

16 Cut these shapes into fifths.

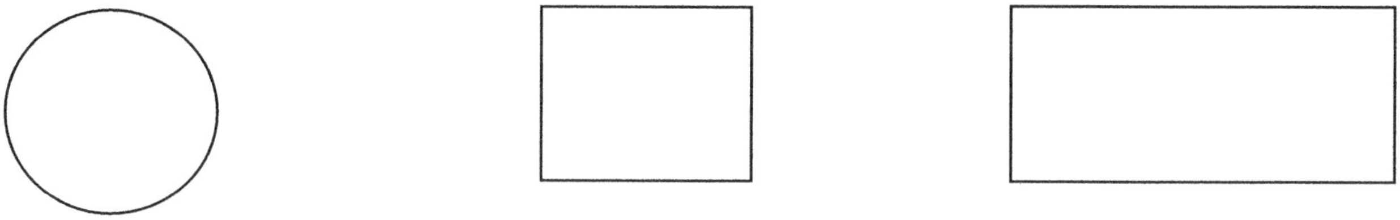

17 Lynn thinks she has cut this shape into fifths. Is she correct? Why?

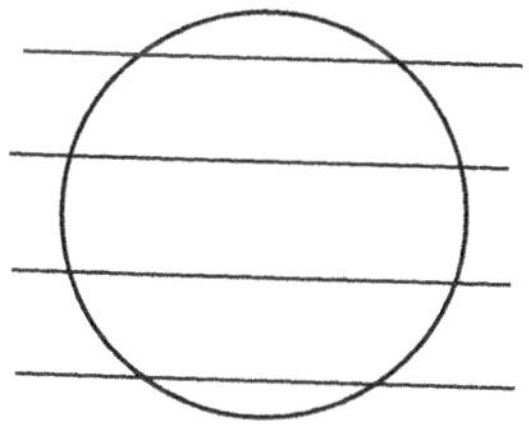

____

____

To work out a fifth, you divide the number by 5.

18 Sue is sharing this chocolate bar into fifths.
How many squares will there be in each fifth?

____

19 Naila made a cake and shared with her family. She cut the cake into equal parts.
She gave her family 4 slices. One slice is left. What fraction of the cake is left?

____

20 Salma made 30 cupcakes for a party. When the party ended, only a fifth of the cupcakes were left.

How many cupcakes were left? ______________________________

One fifth	Two fifths	Three fifths	Four fifths	Five fifths
$\frac{1}{5}$	$\frac{2}{5}$	$\frac{3}{5}$	$\frac{4}{5}$	$\frac{5}{5}$
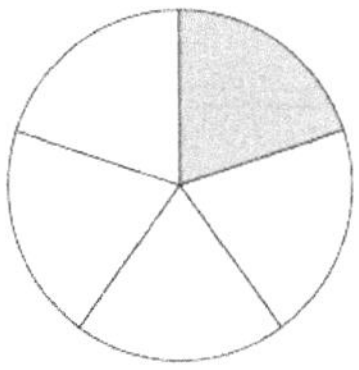	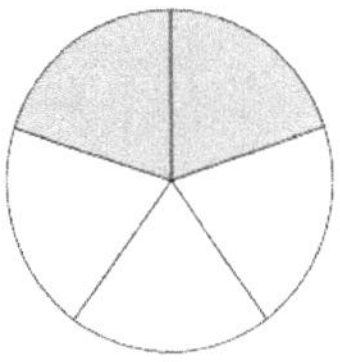	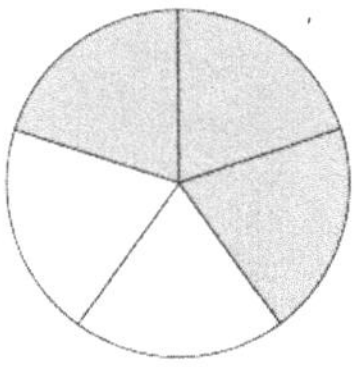	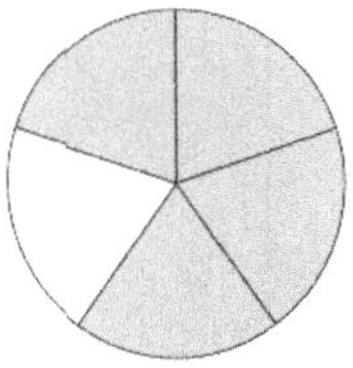	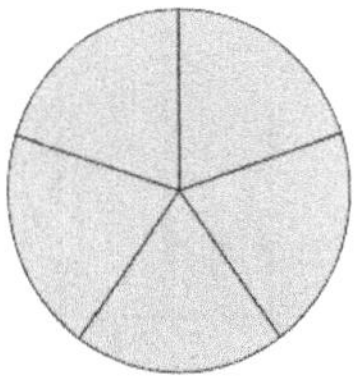
1 out of 5	2 out of 5	3 out of 5	4 out of 5	5 out of 5

21 Shade $\frac{1}{5}$ of this shape.

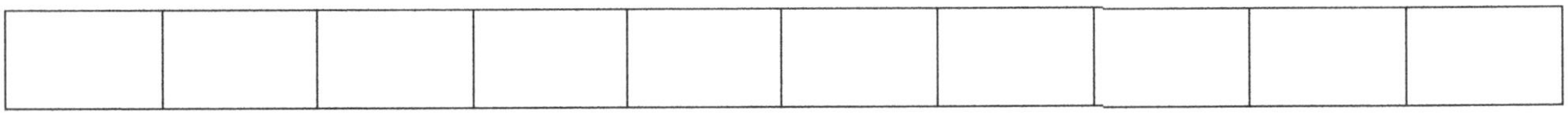

22 Shade $\frac{3}{5}$ of this shape.

23 Compare these fractions using >, < or =

$\frac{3}{5} \bigcirc \frac{1}{5}$	$\frac{1}{4} \bigcirc \frac{1}{5}$	$\frac{5}{5} \bigcirc \frac{4}{4}$

24 Emma is teaching a digital marketing course. Four fifths of the students have found jobs in the industry. What fraction of the students did not find jobs in the industry?

____

25 What fraction of the shapes is coloured in?

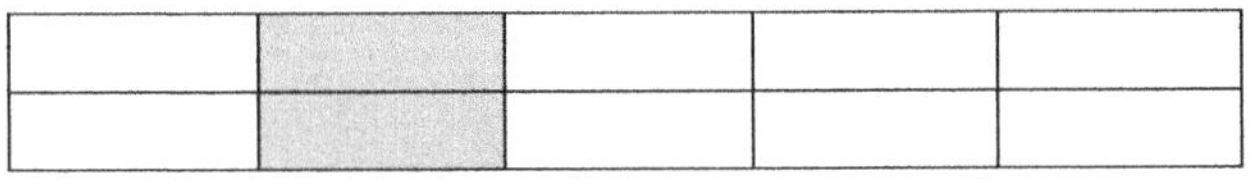

$\frac{\quad}{10}$ or $\frac{\quad}{5}$

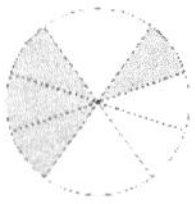

$\frac{\quad}{10}$ or $\frac{\quad}{5}$

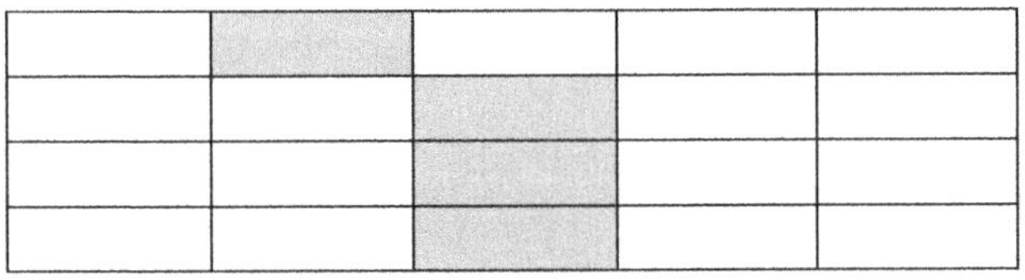

$\frac{\quad}{20}$ or $\frac{\quad}{5}$

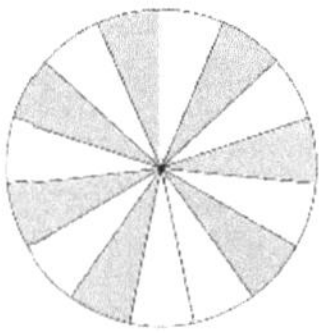

$\frac{\quad}{15}$ or $\frac{\quad}{5}$

If we cut a shape into **10 equal parts** and take 1 part, we've taken **a tenth** of it.

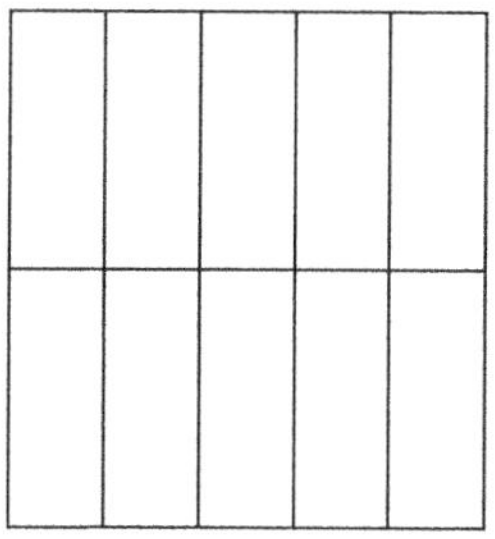

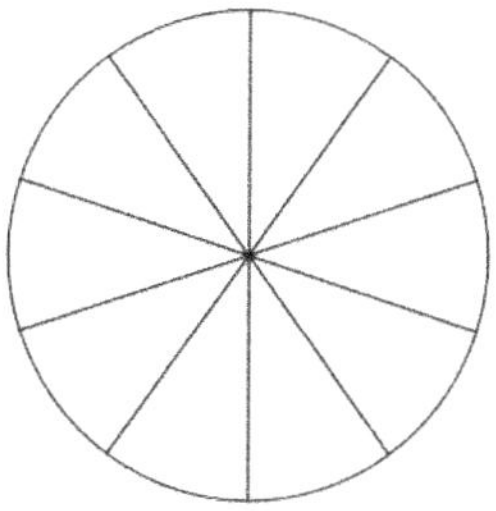

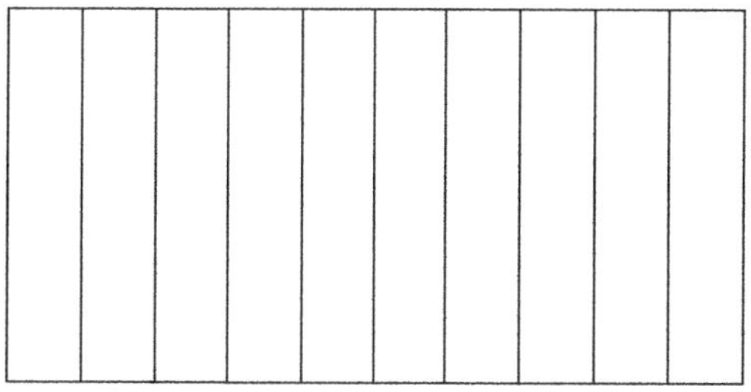

The shape can be cut in different ways, as long as it is split into ten equal parts.

26 Cut these shapes into tenths.

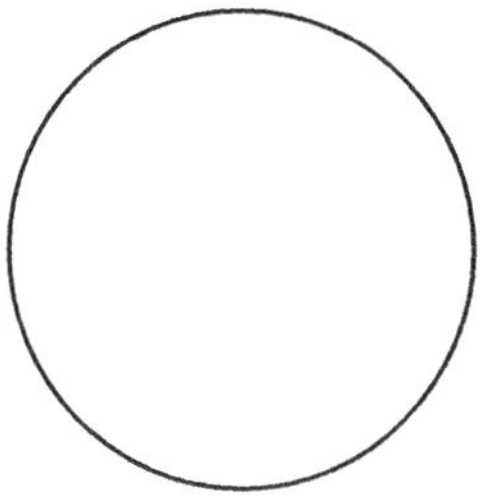

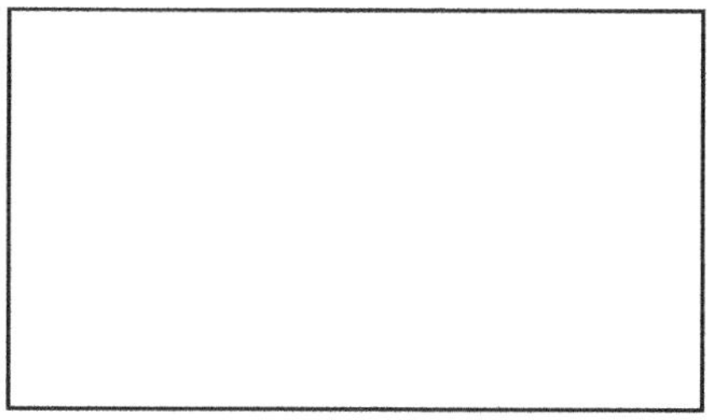

27 Lynn thinks she has cut this shape into tenths.

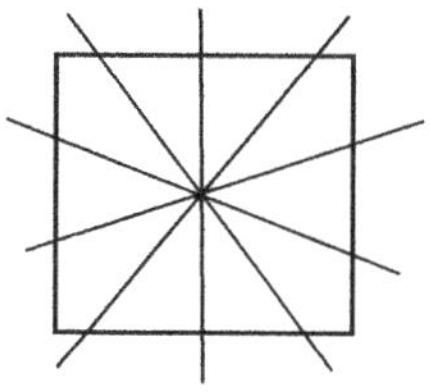

To work out a tenth, you divide the number by 10.

Is she correct? Why? ______________________________

28

a) Nike was given £20.

She will give a tenth of the money to her friend.
How much will she give to her friend?

____

b) Steph has 40 sweets.

She is giving a tenth of them to her friend.
How many sweets is Steph giving to her friend?

____

c) Don will have a tenth of these eggs for breakfast.

How many eggs will Don have for breakfast?

____

d) Yee earns £120 a day. One tenth of that goes towards food.

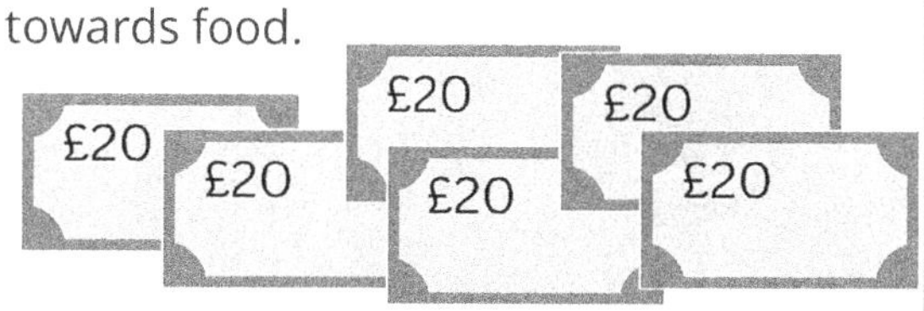

How much does Yee spend on food each day?

____

29 Colour in $\frac{1}{10}$ of this shape.

30 Colour in $\frac{6}{10}$ of this shape.

31 Colour in $\frac{9}{10}$ of this shape.

one tenth	two tenths	three tenths	four tenths	five tenths
$\frac{1}{10}$	$\frac{2}{10}$	$\frac{3}{10}$	$\frac{4}{10}$	$\frac{5}{10}$
	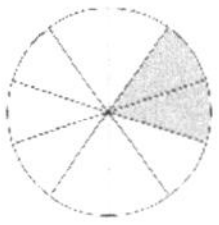	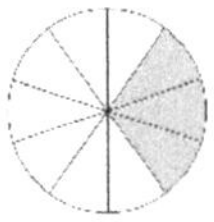	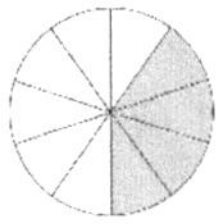	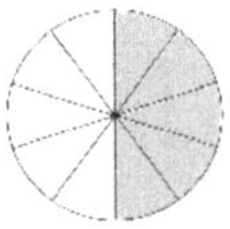
1 out of 10	2 out of 10	3 out of 10	4 out of 10	5 out of 10
six tenth	seven tenths	eight tenths	nine tenths	ten tenths
$\frac{6}{10}$	$\frac{7}{10}$	$\frac{8}{10}$	$\frac{9}{10}$	$\frac{10}{10}$
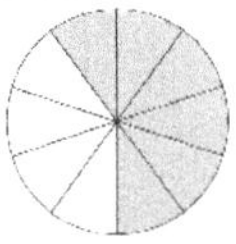	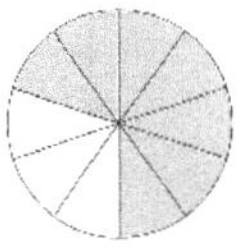	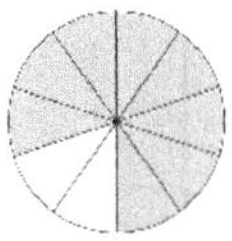	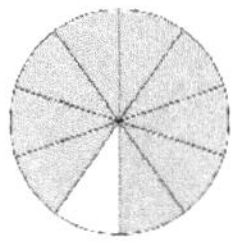	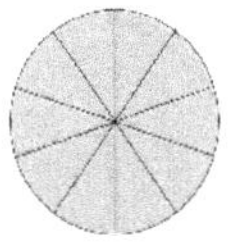
6 out of 10	7 out of 10	8 out of 10	9 out of 10	10 out of 10

32 Compare these fractions using >, < or =

$\frac{2}{10} \bigcirc \frac{8}{10}$	$\frac{2}{5} \bigcirc \frac{1}{10}$	$\frac{1}{4} \bigcirc \frac{2}{10}$

33 What fraction of the whole has been shaded?

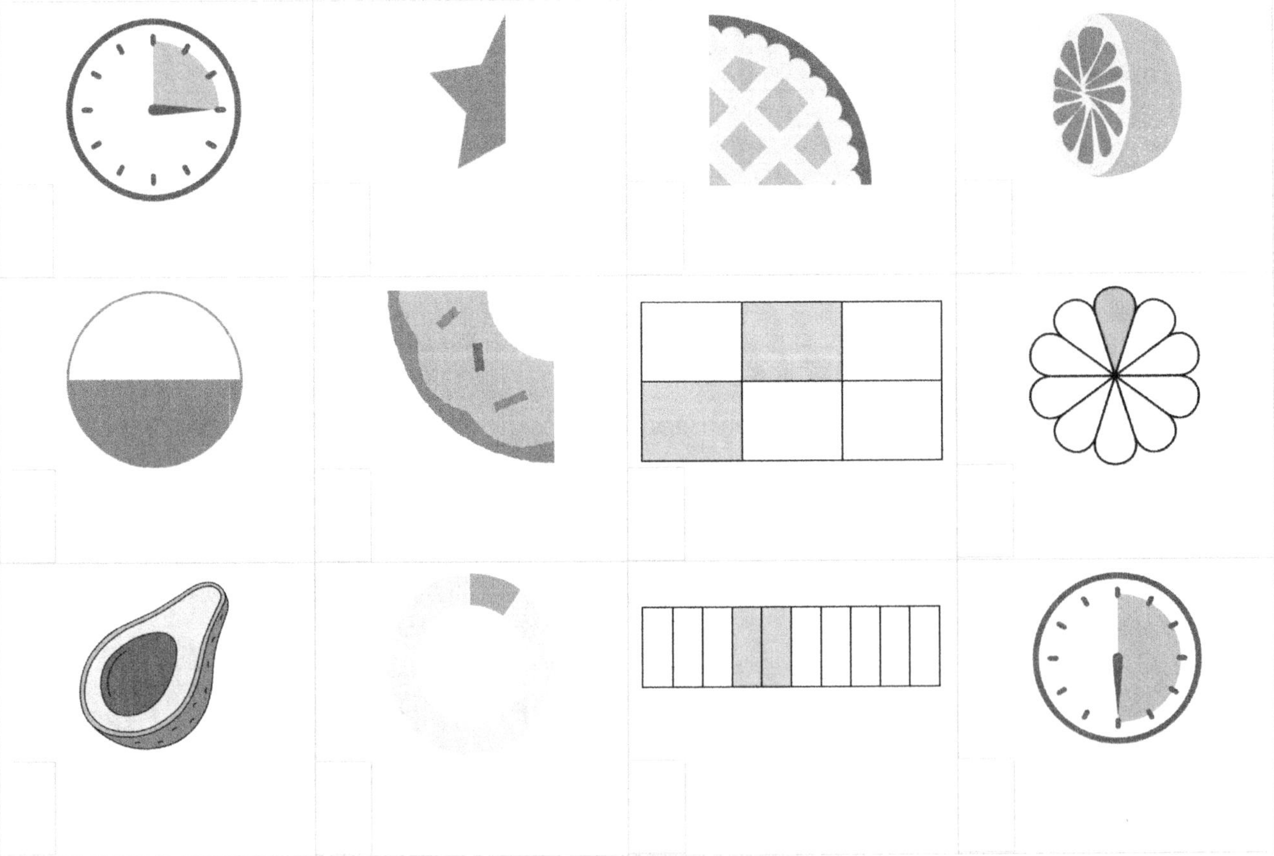

34 Hannah bought these tuna tins.

She thinks a quarter of these is equal to two tins. Is she correct?

____

35 Colour in a third of these rectangles.

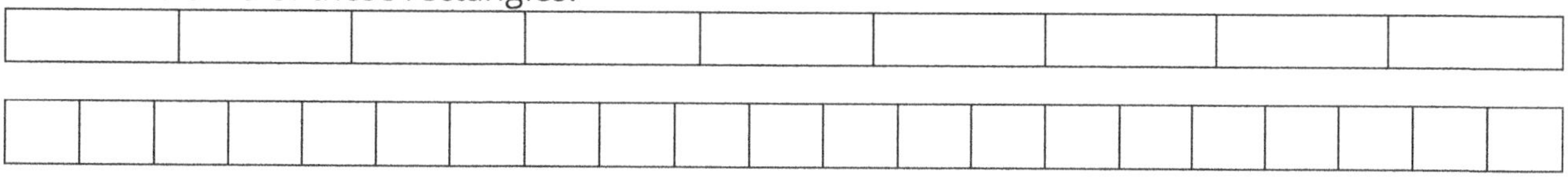

36 Colour in a fifth of these rectangles.

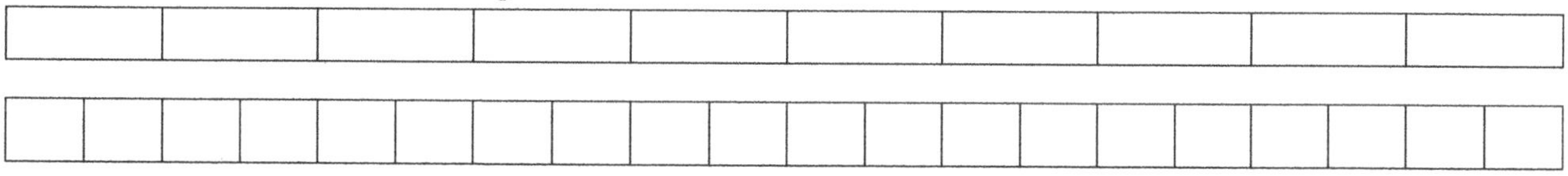

37 Lucy eats $\frac{1}{3}$ of the pizza. What fraction of the pizza is left? ________________

38 a) How many quarters are there in a half? ___ b) How many tenths are there in a fifth? ___

c) How many thirds are there in a whole? ___ d) How many tenths are there in a half? ___

39 a) Circle the fraction that represents the most.

$\frac{1}{3}$ $\frac{1}{10}$ $\frac{1}{4}$ $\frac{1}{2}$ $\frac{1}{5}$

b) Put the fractions in order, starting with the one that represents the most.

__ __ __

c) Circle the fraction that represents the least.

$\frac{1}{2}$ $\frac{4}{10}$ $\frac{1}{4}$ $\frac{3}{3}$ $\frac{1}{5}$

d) Put the fractions in order, starting with the one that represents the most.

__ __ __ __ __

40 An album has 12 songs. If you like three of them, what fraction of the songs do you like? __________

41 Roi says: 'To find a quarter of a number, you halve and halve the number again'. Is Roy correct?

____

42 Jane designs a webpage. The page has a width of 40 cm. An image will take $\frac{1}{4}$ of the width of the page. Jane says the image needs to be 11 cm in width. Is Jane correct? Show why you think this.

____

43 Nine out of 15 boxes have been shipped. What fraction of the boxes has been shipped?

____

44 A customer buys a sofa which costs £998. They must pay $\frac{1}{2}$ of the price today. They will pay the rest when the sofa is delivered. How much will the customer pay today?

____

45 Zoe plans to buy a bike costing £400. The store is offering $\frac{1}{4}$ off the price. How much will Zoe pay if she takes advantage of the discount?

____

46 A bakery made 30 loaves of bread today. Ten of the loaves are whole wheat. What fraction of the loaves are whole wheat? _______________

47 Desi bought a fridge for £500. She will pay this amount over ten instalments.

a) How much will she pay each month? _______________

b) What fraction of the whole amount is left to pay, after the first instalment is paid? _______________

48 A school has 15 classes. $\frac{1}{5}$ of all classes go on a field trip. How many classes remain?

____

49 Layla has a ribbon that is 12 meters long. She uses $\frac{1}{3}$ of the ribbon. How much ribbon is left?

____

50 A recipe needs three quarters of a cup of oil but you only have half a cup. How much more oil do you need? Express your answer as a fraction.

____

Progress check

1 Cut these shapes into thirds.

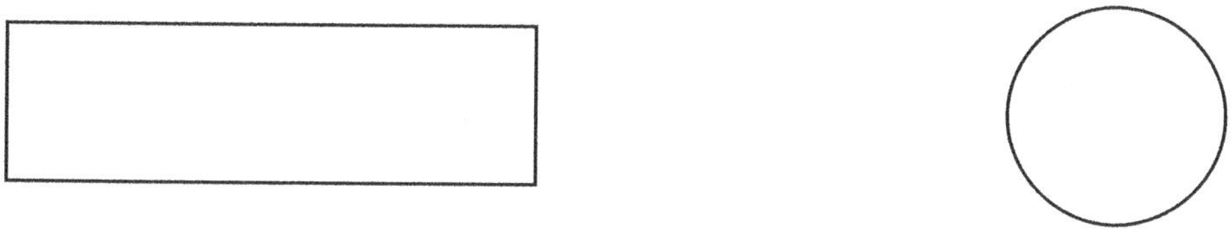

2 Jade represents $\frac{1}{3}$ like this:

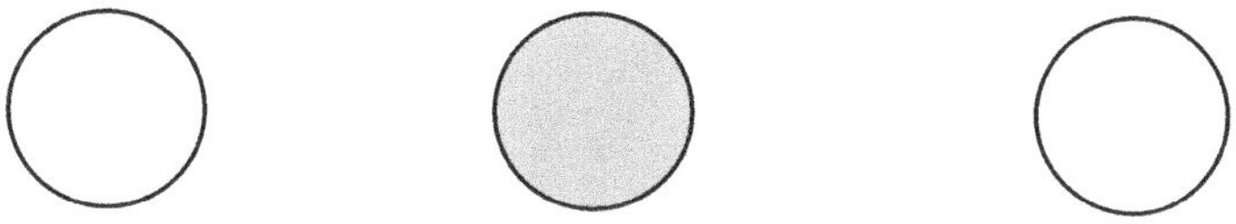

Is she correct? Explain your answer.

____

3 Fill in the gaps below:

Fraction in digits	Fraction in words
	three quarters
$\frac{2}{5}$	
	six tenths

4 Which of these is greater: $\frac{1}{4}$ of £84 or $\frac{1}{5}$ of £100?

____

5 Fill in the gap below:

$$\frac{1}{10} = \frac{7}{}$$

6 Ayla says: '$\frac{1}{10}$ is bigger than $\frac{1}{3}$ because 10 is bigger than 3.' Is Ayla correct? Explain your answer.

____

7 Ethan is baking cookies. The recipe he is using is for four people and asks for 120g of sugar. Ethan is making cookies just for one person. How much sugar will he need?

____

Read, write and use decimals up to two decimal places

We use decimals when:

- working with money
- working with measures
- converting fractions into decimals
- working with time
- measuring body weight
- cooking

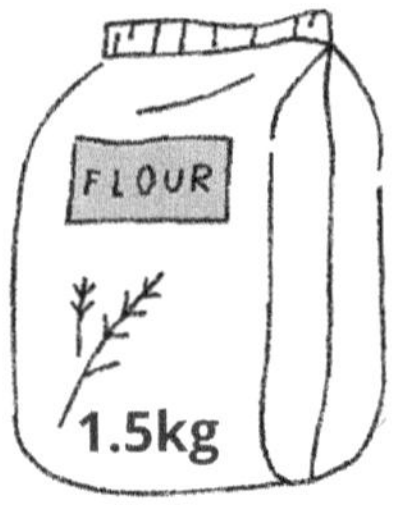

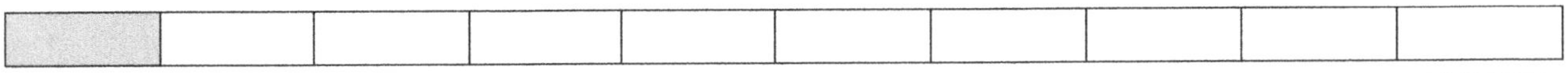

$\frac{1}{10}$ means 1 out of 10. As a decimal, this is 0.1

We read this as **zero point one**.

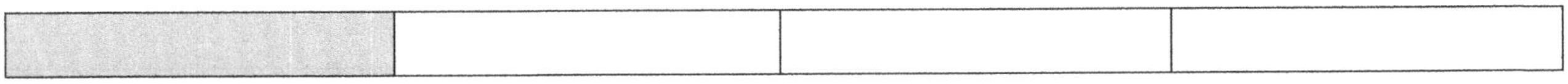

$\frac{1}{4}$ means 1 out of 4. As a decimal, this is 0.25

We read this as **zero point two five**.

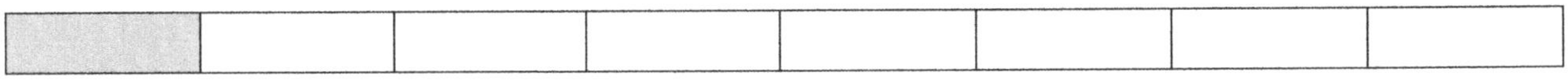

$\frac{1}{8}$ means 1 out of 8. As a decimal, this is 0.125

We read this as **zero point one two five**.
We read digits separately after the decimal point.

1 Match the decimals in digits with how you say them in words.

3.40	16.7	0.95	10.02	140.3

sixteen point seven	ten point zero two	one hundred and forty point three	three point four zero	zero point nine five

Representing decimals

Here is how £11.11 could be represented. Decimals have a **decimal point**.

tens	units		tenths	hundredths
1	**1**	.	**1**	**1**

The digits after the decimal point are called **decimal places**.

This is another way to represent 11.11

tens	units		tenths	hundredths
1	**1**	**.**	**1**	**1**
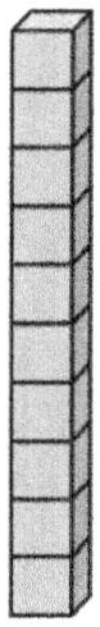				

A **unit** () is 10 times smaller than a block of ten (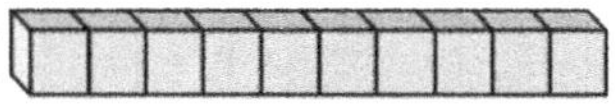)

A **tenth** () is 10 times smaller than a unit ().

A **hundredth** () is 10 times smaller than a tenth ().

This is how money links to decimals. 1 represents £1.

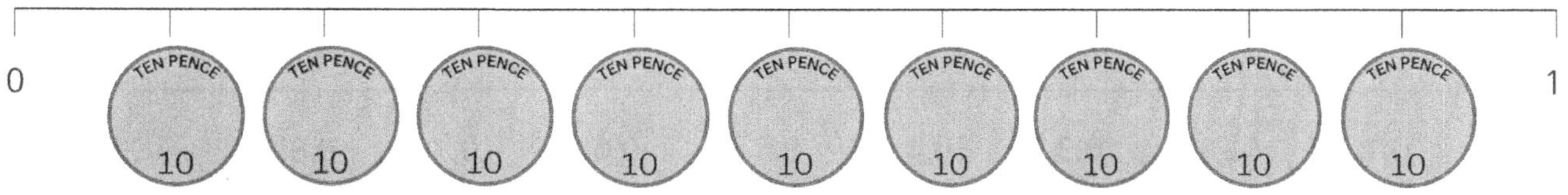

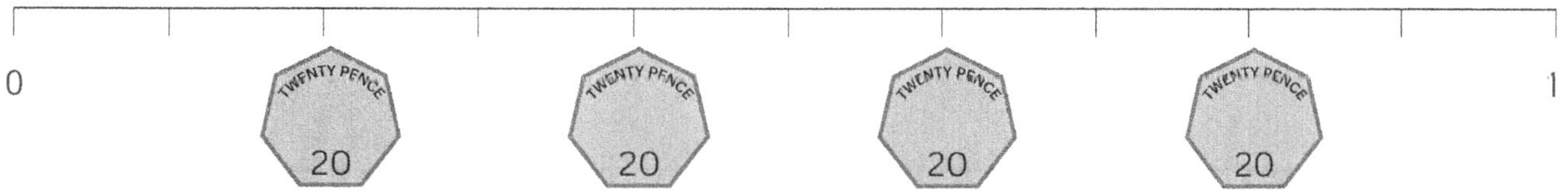

0.5 represents the same amount as 0.50
0.2 represents the same amount as 0.20
0.9 represents the same amount as 0.90
1 represents the same amount as 1.0

2 Write down what decimals these represent. The first one has been done for you.

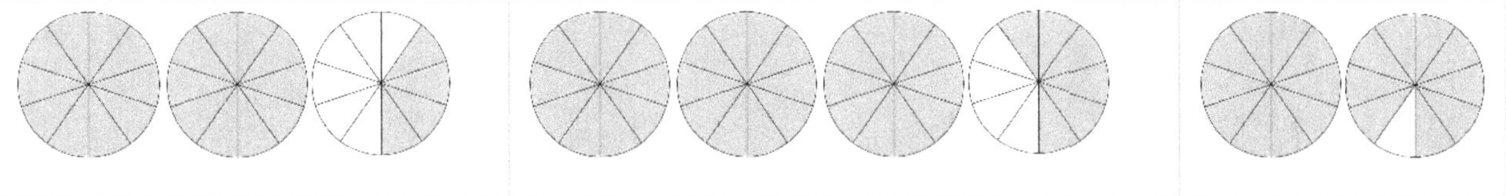

2.4		

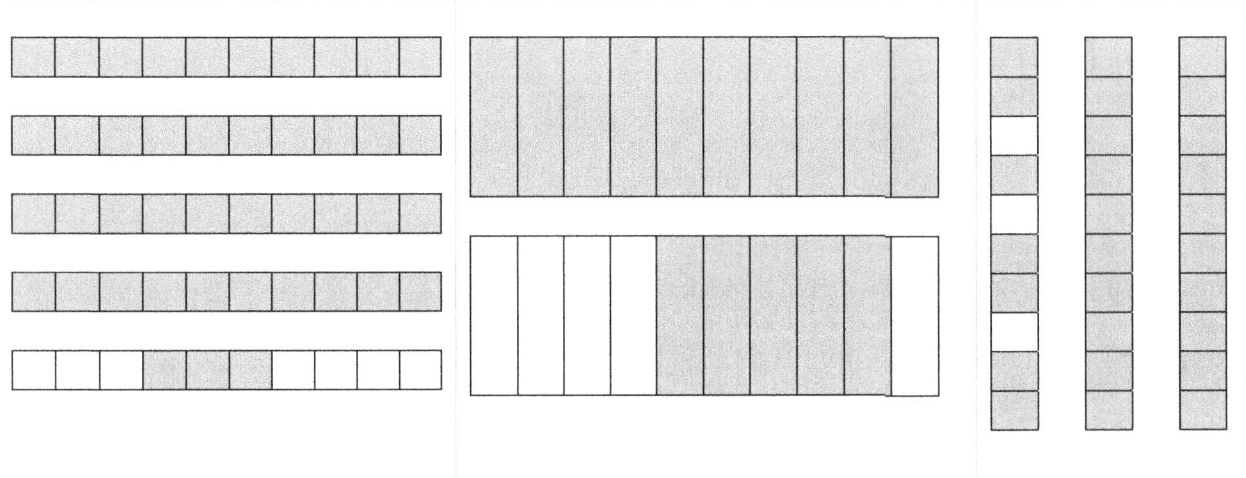

4.3		

This number line starts with 0 and goes up to 1 (a whole).

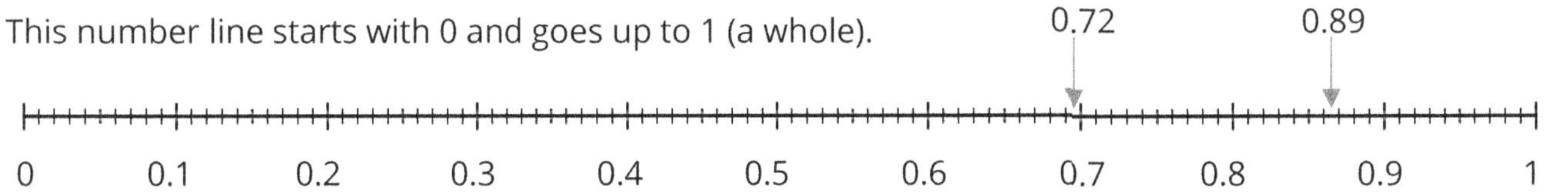

3 Write the decimal that represents what the arrow is pointing at.

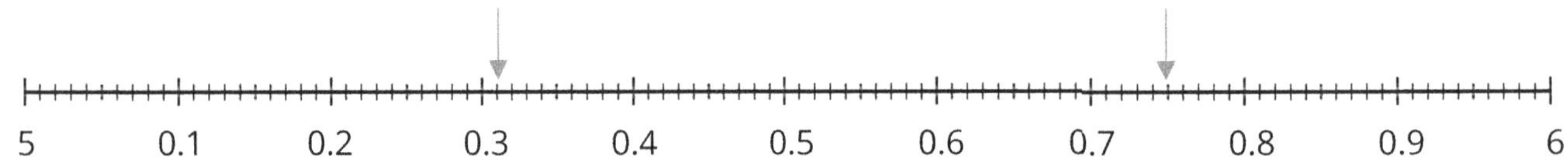

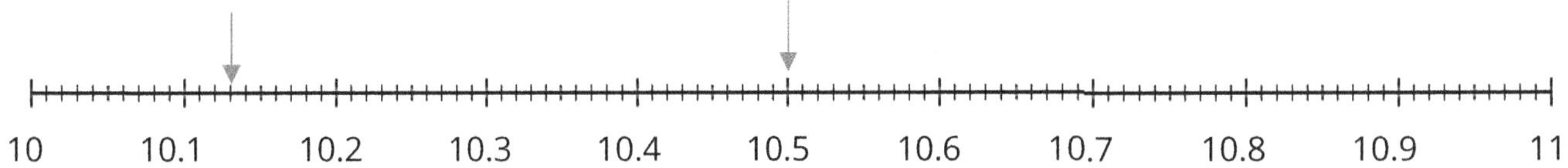

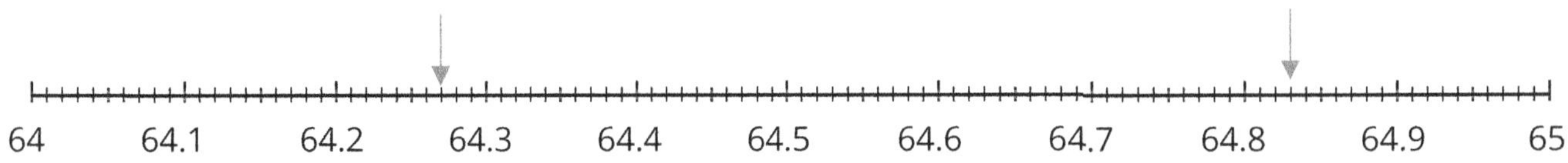

Linking decimals with measurements

4 Fill the gaps.

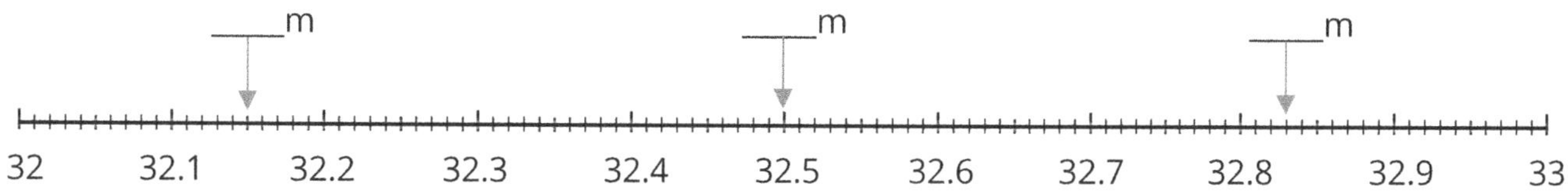

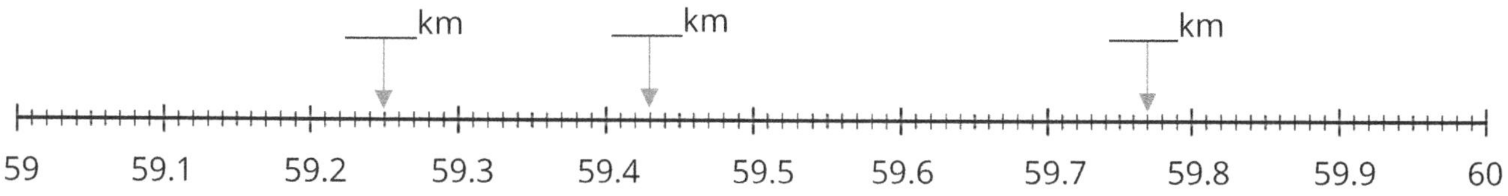

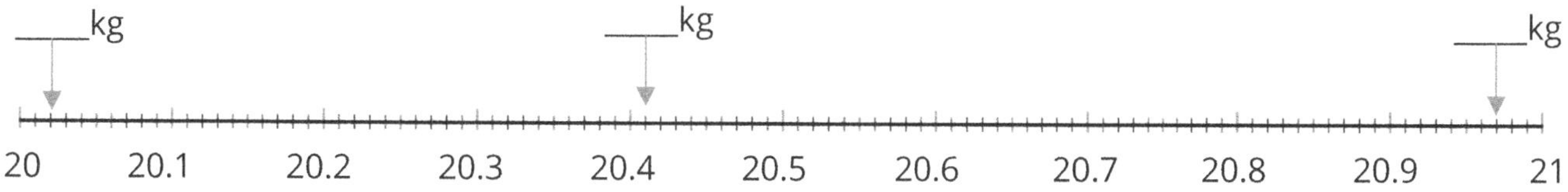

5 Read these scales.

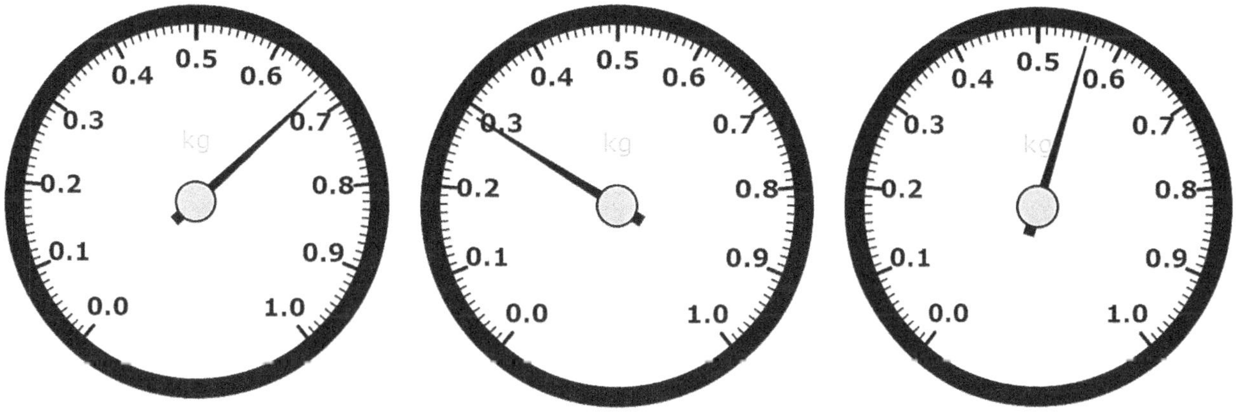

_____________________ _____________________ _____________________

6 a) What is the value of the digit 6 in the decimal 3.67? _____________________

b) In the number 8.09, identify the hundredths place. _____________________

c) Which is greater: 5.67 or 5.76? _____________________

7 Put these numbers in order, starting with the smallest.

0.75 1.08 1.80 0.80

________, ________, ________, ________

Adding decimals

	2.5		0.8 (carry 1)		0.63 (carry 1, 1)		6.85 (carry 1)		3.8 (carry 1)
+	3.2	+	5.4	+	0.79	+	2.9	+	5.57
	5.7		6.2		1.42		9.75		9.37

8 Add these decimals.

6.7 + 3.1	5.5 + 4.2	5.6 + 2.4	5.0 + 3.7	7.34 + 2.37	11.39 + 8.67	1.46 + 5.78
4.84 + 4.17	7.64 + 1.76	1.55 + 7.29	8.49 + 2.6	2.66 + 7.78	9.1 + 3.86	6.9 + 7.49

When adding or subtracting decimals, we line them up so that the decimal point goes under the decimal point, the units under the units, the tenths, and then we calculate as if they were whole numbers.

Subtracting decimals

	7.5		5.8 (5 → 4, 18)		9.73 (9 → 8, 7 → 6, 13)		6.8 (6 → 5, 8 → 7, 10)		5.0 (5 → 4, 10 → 9, 1)
-	3.2	-	2.9	-	0.99	-	2.96	-	4.16
	4.3		2.9		8.74		3.84		0.84

9 Subtract these decimals.

6.7 - 3.1	5.5 - 4.4	8.6 - 5.4	9.0 - 8.7	9.35 - 7.36	9.37 - 4.68	5.74 - 4.59
8.1 - 4.84	7.63 - 2.7	7.2 - 5.16	8.3 - 2.67	7.0 - 2.7	9.0 - 3.84	7.4 - 6.84

10 a) Luis is a postman. He is putting these boxes in order of weight, starting with the lightest. Put them in order.

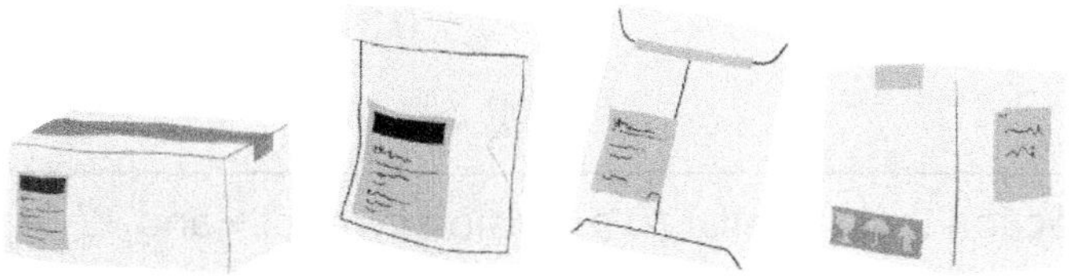

A)10.6kg B)2.45kg C)1.92kg D) 10.0kg

_____, _____, _____, _____

b) A bag of sugar weighs 1kg. A bag of flour weighs 1.5kg.

How much do they weigh altogether?

____

c) Joss works in different offices each day. These are the distances to each office, from her house.

Monday **5.35km**

Tuesday **4.8km**

Wednesday **5.48km**

Thursday **5.0km**

Friday **4.65km**

Which day does she travel the shortest distance?

____

d) A water bottle holds 0.75*l*. How much do two bottles hold?

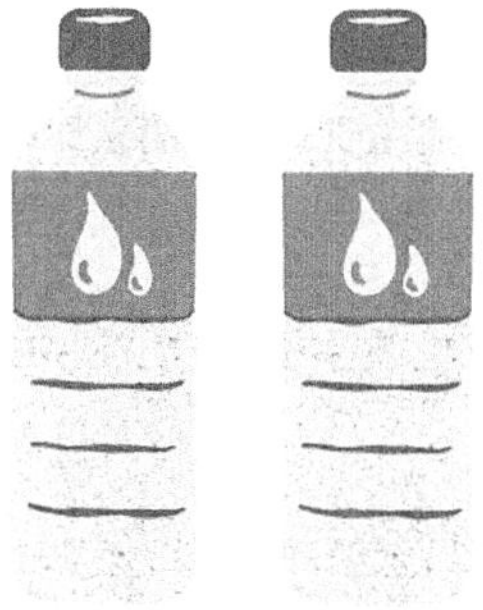

____

e) Laura needs a new wardrobe. The space where the wardrobe will be put is 2.5m.

These are the lengths of the wardrobes Laura likes.

A	**B**	**C**	**D**
2.30m	2.05m	2.92m	2.69m

Laura wants the longest wardrobe that fits in the space. Which wardrobe should Laura buy?

____

f) Ola has 0.75m of ribbon. She uses 0.4m to decorate a cake.

How much ribbon is left?

____

11 a) Circle the greatest decimal: 0.6 0.5 0.11

b) Circle the greatest decimal: 0.67 0.5 0.15

c) Circle the greatest number: 0.6 0.59 1.1

d) Circle the greatest number: 10.9 20.3 15. 11

e) Circle the greatest number: 1.0 0.5 0.1

> Look at whole number first.
> The decimal with the greater whole number is the greatest.
> Example: 3.6 >2.6
> Then look at the first digit after the decimal point: 3.6 >3.5
> Then look at the second digit after the decimal point: 3.63 >3.62

Decimals represent part of a whole, just like fractions.

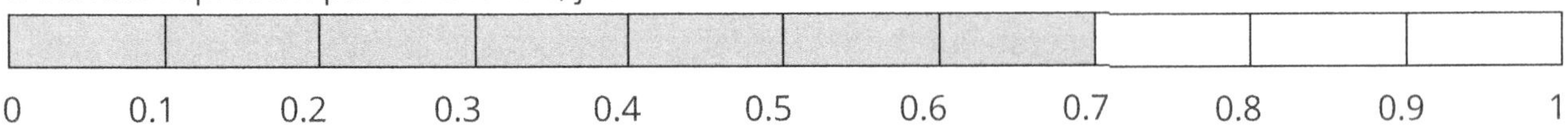

Use a calculator to work out:

$\frac{1}{2}$ by dividing 1 by 2

$\frac{1}{10}$ by dividing 1 by 10

A fraction looks similar to a division symbol ÷ and it means division.

$\frac{1}{2}$ means 1 divided by 2.

$\frac{1}{10}$ means 1 divided by 2.

12 Match each fraction with the decimal it is equal to.

$\frac{1}{4}$	$\frac{1}{5}$	$\frac{1}{10}$	$\frac{3}{4}$	$\frac{1}{3}$

0.2	0.75	0.25	0.33	0.1

13 Which of these is the smallest number? Circle it.

467.42 467.14 467.24 467.41

14 Which of these is the highest number? Circle it.

823.6 823.56 823.8 823.66

15 John measures the weight of some parcels on a digital scale.

A. 12.1 kg B. 12.11 kg C. 12.01 kg D. 12.21 kg

Put these measures in order, starting with the lightest: _________, _________, _________, _________

16 Mike decides to keep a running total of all the miles he cycles each week. Complete the sequence to show his running totals for week 4: Running Total: 11.5, 23, 34.5, ____

17 A recipe requires 0.75 kg of flour and 0.25 kg of sugar. Write down the total weight of these ingredients.

____

Progress check

1 Represent £11.11 by drawing coins and notes.

2 The cost of a bag of sugar is £0.62 and the cost of a bag of flour is one and a half pounds. How much do the two items cost altogether?

____

3 The length of a wood plank is 16.6cm.

a) What is the length of two planks of wood put next to each other?

____

b) Would the two pieces put together be long enough to cover a strip of the floor that is 30cm long?

____

4 Fill in the gaps.

Decimal in digits	Decimal in words
	zero point six seven
14.09	
	four hundred point nine

5 What is the same about these two numbers and what is different? 3.75 and 3.57

____

6 Here are three digit cards: 4 5 7

a) Make every decimal you can with them.

____

b) Write them in order, starting with the smallest.

____

Recognise and continue sequences that involve decimals

We come across decimal sequences in:

- shoe sizes
- money
- measurement
- time

Money can go up in a sequence. Example: £0.50, £1.00, £1.50, £2.00, £2.50, £3.00, ...
Shoe sizes go up in a sequence. Example: 4, 4.5, 5, 5.5, 6, 6.5, 7, 7.5, 8, 8.5, ...

When counting forward (e.g. 0.2, 0.4, 0.6, 0.8) or backward (e.g. 0.8, 0.6, 0.4, 0.2) by the same amount each time, we are creating a sequence.

In this number line, numbers are going up by 0.5, so 0.5 is added each time.

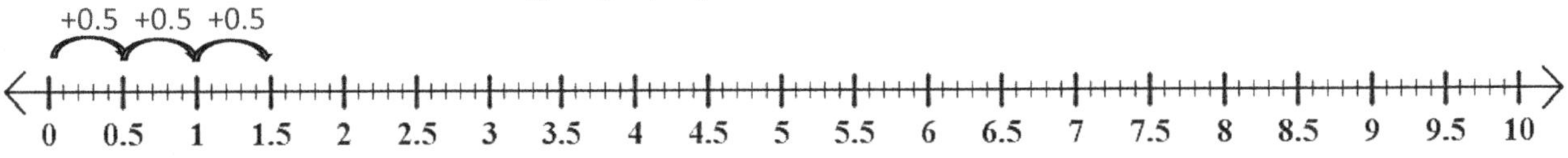

In this number line, numbers are going up by 0.25, so 0.25 is added each time.

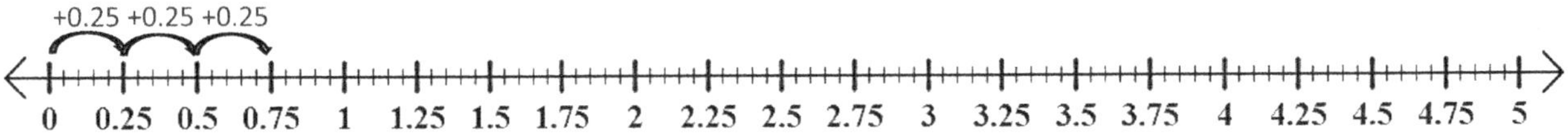

To work out what the sequence is going up or down by, you find the difference between two numbers next to each other.
For this sequence: 3.2, 4.4, 5.6, 6.8, 8.0 we do: 4.4 - 3.2 = 1.2 or 5.6 - 4.4 = 1.2 so the sequence increases by 1.2

For this sequence: 1.4, 1.1, 0.8, 0.5, 0.2 we do: 1.4 - 1.1 = 0.3 or 0.8 - 0.5 = 0.3 so the sequence decreases by 0.3

1 Anna is building a bench in her garden. She needs five planks of wood that measure 1.2m.
She counts: 1.2, 2.4, 3.6, _____, _____
Add the next two decimals to the sequence.

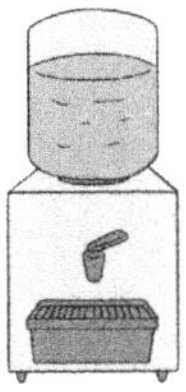

2 Lina is filling a water dispenser with water bottles. Each water bottle holds 0.75 litres.
She keeps count of the number of litres she has put already in: 0.75, 1.5, 2.25, _____, _____, _____
a) Add the next three decimals to the sequence.
b) The water dispenser holds 6 litres of water. How many more bottles does Lina need to empty into the dispenser?

____

3 Nesa is buying a water bottle from the vending machine. She has 10p coins only.
She counts as she puts the money into the machine: 0.1, 0.2, 0.3, 0.4, _____, _____
After 0.9 she says 0.10. Is she correct? Why?

____

> Remember, in decimal numbers, 0.1 is the same as 0.10 - adding a zero at the end doesn't change its value!

4 Nita is counting her change. She has £2.40 in 20p coins. As she takes a coin each time, she works out how much is left: 2.40, 2.20, 2.00, _____, _____, _____
Continue the sequence.

5 Karen works in a bakery. She knows she has 5kg of butter in the fridge. Each time she takes out a pack, she works out how much is left in the fridge: 5, 4.75, 4.5, _____, _____

a) Continue the sequence.

b) How much does a pack of butter weigh?

____

c) How many packs were there in total?

____

6 Continue these sequences.

0.2, 0.4, 0.6, 0.8, _____, _____	0.22, 0.32, 0.42, _____, _____	10.52, 10.57, _____, _____
3.7, 3.5, 3.3, _____, _____	1.66, 1.55, 1.44, _____, _____	9.50, 9.56, 9.62, _____, _____
7.4, 7.7, 8.0, _____, _____	6.4, 7.2, 8.0, _____, _____	0.8, 1.6, 2.4, _____, _____

7 Fill the gaps in these sequences.

5.5, 5.6, 5.7, _____, _____, _____	1.4, 2.1, 2.8, _____, _____, 4.9
_____, _____, 3.0, _____, _____ 3.3	6.6, 5.5, 4.4, _____, _____, 1.1
4.5, 4.7, 4.9, _____, _____, 5.5	5.34, 5.38, 5.42, _____, 5.60, _____, _____

8 Women shoe sizes go up by 0.5. Suki has tried size 6. She wants one size smaller. What size is that?

____

9 A sequence goes like this: 1.2, 1.8, 2.4, 3.6 .... Will 7.0 be in this sequence? Show your working.

____

10 A sequence goes like this: 2.8, 2.5, 2.2, ... What comes before 2.8?

____

11 A sequence goes like this: 4.9, 5.8, 6.7, ... Bardha says: 'This is as easy as going up by one with the whole numbers and down by one with the tenths'.

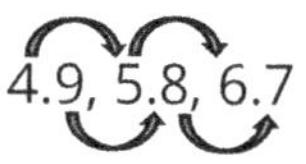

Do you agree? Explain. ____

12 David wants to install some shelves in his living room. These are the heights of the shelves in metres: 1.0 m, 1.35 m, 1.7 m. The heights follow a pattern. What height will the next shelf be?

____

13 Potted plants can be bought in four different heights. The heights follow a sequence. What are the next two heights? 0.3 m, 0.5 m, 0.7 m, _____ m, _____ m

14 Sam prepares dough for different loaves of bread. The amount of dough increases in sequence. What is the next number in the sequence? 0.4 kg, 0.8 kg, 1.2 kg, _____ kg

15 Carla is measuring the growth of her plant. These are the heights of her plant in centimetres over weeks: 2.0 cm, 2.5 cm, 3.0 cm. The growth follows a pattern. Carla wants to predict the height after the next week. What height will the plant be?

____

16 Tom decides to keep a running total of all the pages he reads each week. Complete the sequence to show his running totals for week 4. Running Total: 20.5, 41, 61.5, _____

17 Look at this sequence and identify the next number: 2.5, 3.0, 3.5, ...

A. 3.10	B. 4.0	C. 4.5	D. 5.0

18 Look at this sequence: 1.1, 1.2, 1.3, ... What is the difference between the numbers?

A. 0.01	B. 0.1	C. 1	D. 10

19 This sequence has an error: 4.0, 4.2, 4.4, 4.7, ... Rewrite it in the correct way: ______________________

Progress check

1 Fill the gaps in these decimal sequences:

3.3, 3.4, _____, 3.6, _____, 3.8	6.1, 6.4, _____, 7.0, _____	_____, 2.4, 3.6, _____, _____, _____

2 Sunny is asked to continue this sequence: 2.4, 3.6, 4.8 , _____, _____
His answer is: 2.4, 3.6, 4.8 , 5.10 , 6.12. Is he correct? Explain your answer.

____

3 Guillermo says 'In a sequence, you always need to add or take away the same amount". Is he correct? Explain.

____

Calculate with money using decimal notation and express money correctly in pounds and pence

We use money when:

- shopping
- paying bills
- counting savings
- taking up loans
- borrowing or lending money

A picture frame costs £11.23. That is 11 pounds and 23 pence.

The **decimal point** separates the pounds from the pence.

hundreds	tens	units	.	tenths	hundredths
	1	1	.	2	3
				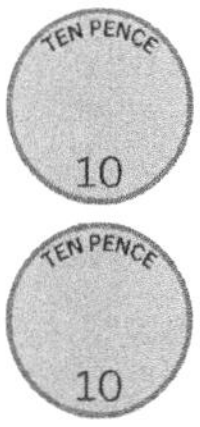	

A **unit** (pound coin) is 10 times smaller than a block of ten (£10 note).

A **tenth** (10p coin) is 10 times smaller than a unit (pound coin).

A **hundredth** (1p coin) is 10 times smaller than a tenth (10p coin).

1 £21.00 has been added to the table. Write these amounts: £135. 05 £405.89 £0.05

hundreds	tens	units	.	tenths	hundredths
	2	1		0	0

50p = £0.50 74p = £0.74 99p = £0.99 5p = £0.05 31p = £0.31 65p = £0.65 8p = £0.08

2 Write the value in pence of these amounts.

£0.77	£0.80	£0.08	£0.70	£0.99	£0.09	£0.90	£0.50	£0.05
___p	___p	___p	___p	___p	___p	___p	___p	___p

3 Draw the coin or note that has the same value as these.

=

=

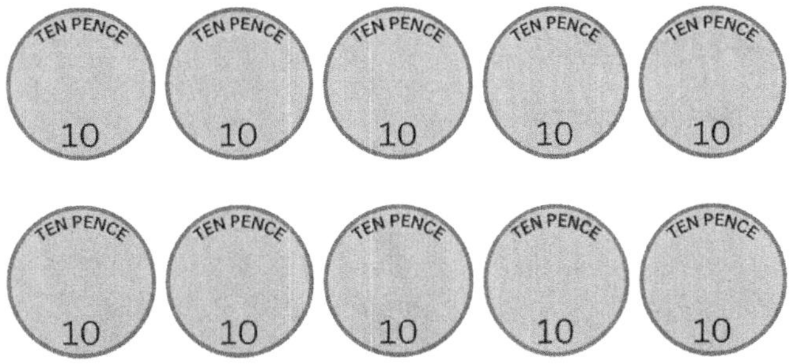

=

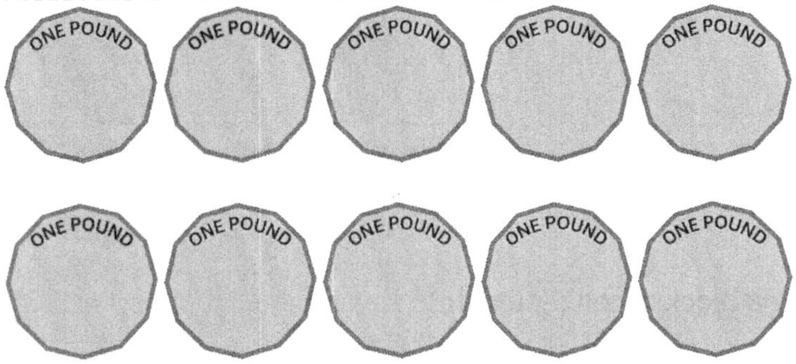

=

4 a) What are one hundred 1p coins worth? _____ b) What are two hundred 1p coins worth? _____

5 Write the value in pence of these amounts.

£1.77	£3.80	£4.08	£1.70	£5.99	£5.09	£2.90	£9.50	£2.05
___p	___p	___p	___p	___p	___p	___p	___p	___p

6 These are the coins and notes that Ken has in his wallet.

How much money does he have in his wallet? ______________________________

Adding money

Example 1

11.62 + 0.40

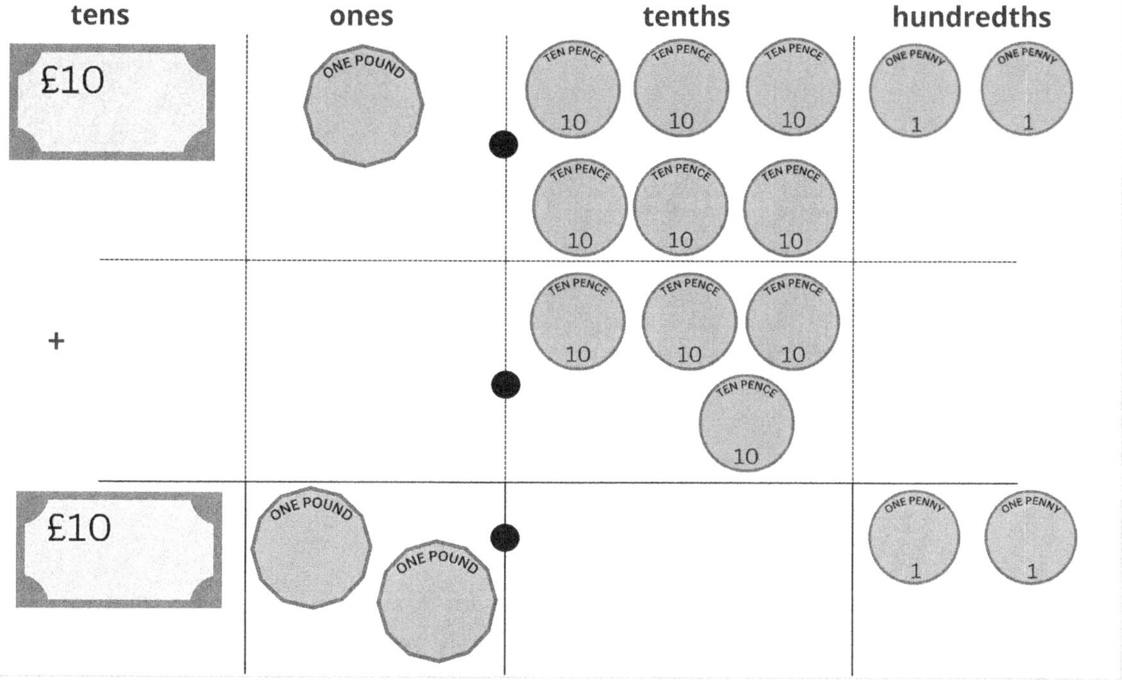

		1			
		0	•	4	0
+	1	1	•	6	2
£	1	2	•	0	4

2 + 0 = 2
As 6 + 4 = 10, we carry 10 over to the ones column and put 0 in the tenths column. Then 0 + 1 + 1 = 2 and 0 + 1 = 1

Example 2

			1		1	
		3	0	•	8	6
+	1	0	7	•	5	4
£	1	3	8	•	4	0

As 6 + 4 = 10, we carry 10 over to the tenths column and put 0 in the hundredths column.
As 8 + 5 = 1 = 14, we carry 10 over to the ones column and put 4 in the tenths column. Then 0 + 7 + 1= 8 and 3 + 0 = 3 and 0 + 1 = 1

Example 3

	1	1			1	
	5	6	7	•	0	2
+	1	7	4	•	2	9
£	7	4	1	•	3	1

As 2 + 9 = 11, we carry 10 over to the tenths column and put 1 in the hundredths column. Then 0 + 2 + 1 = 3.
As 7 + 4 = 11, we carry 10 over to the tens column and put 1 in the ones column.
As 6 + 7 + 1 = 14, we carry 10 to the hundreds column and put 4 in the tens column. Then 5 + 1 + 1 = 7.

Example 4

	1	1				
	5	6	9	•	3	4
+	3	8	7	•	5	1
£	9	5	6	•	8	5

4 + 1 = 5 and 3 + 5 = 8
As 9 + 7 = 16, we carry 10 over to the tens column and put 6 in the ones column.
As 6 + 8 + 1 = 15, we carry 10 to the hundreds column and put 5 in the tens column. Then 5 + 3 + 1 = 9

7 Fill the gaps.

403.87 + 339.08	112 + 78.90	186.40 + 476.54	530.80 + 74.65	376.46 + 533.73	197.35 + 456.65	363.54 + 365.45
463.80 + 374.51	374.66 + 46.27	105.36 + 0.93	436.44 + 6.57	583.72 + 37.46	318.26 + 372.98	473.64 + 437.37
378.47 + 475.83	350.47 + 437.83	376.44 + 400.00	362.46 + 75.36	374.55 + 47.46	100 + 47.69	365.48 + 357.67

8 Gloria buys movie tickets. She buys two tickets for £6.50 each and two popcorns for £4.20 each. What is the total cost?

____

9 Cala is buying a jumper and a hat.
The hat costs £23.99 and the jumper £37.99.
How much do the items cost altogether?

____

____

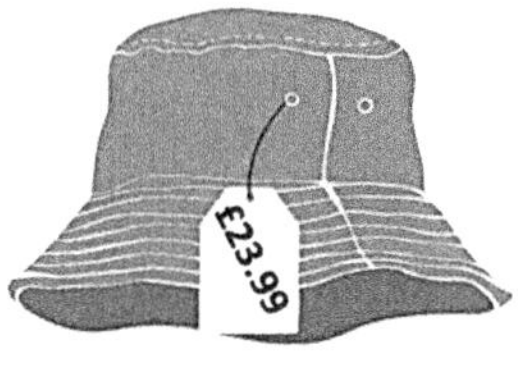

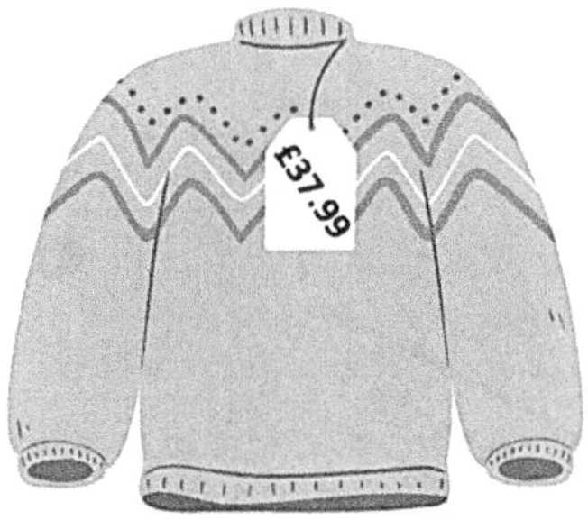

10 Janet buys a new book for £12.99 and a pen for £1.85. How much does she spend in total?

____

11 This protein bar costs 39p. Gary is buying two of them. How much will he pay? Give your answer in pounds.

____

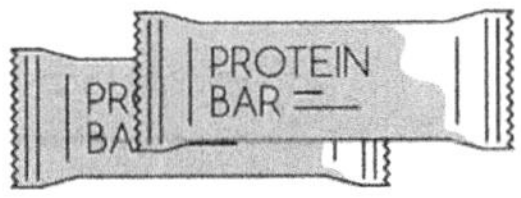

12 A store had sales of £300.95 in the morning. The store had sales of £275.35 in the afternoon. What were the total sales?

____

13 Work out the difference between £15.50 and £12.75. ______________________________

Subtracting money

Example 1

11.52 - 9.90

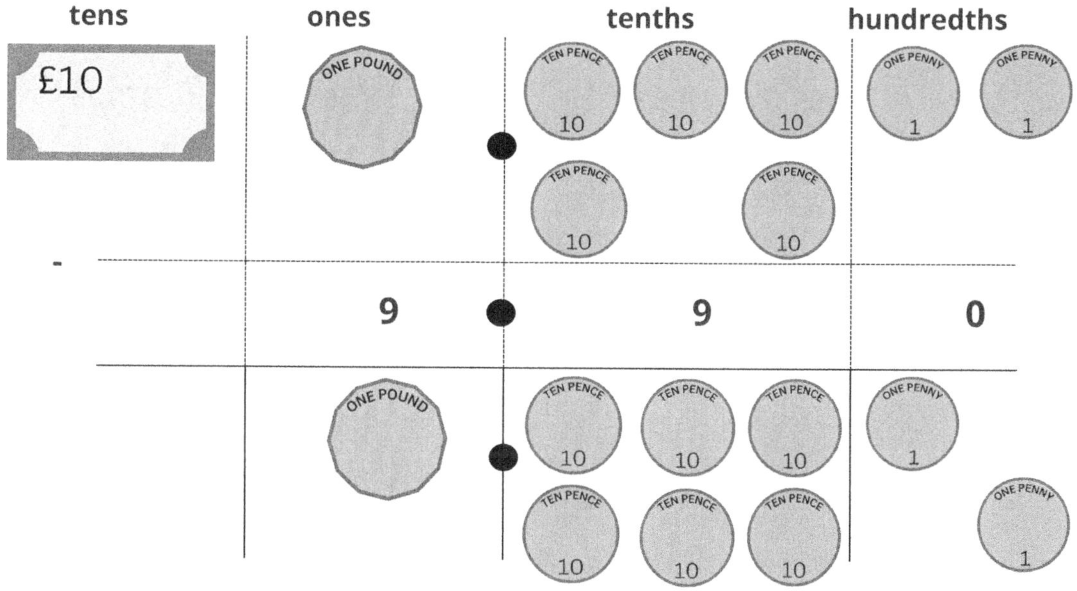

	1	1 (0)	• (1)5	2
-		9	• 9	0
£		1	• 6	2

2 – 0 = 2

As we can't take away 9 from 5, we borrow a pound, which has 10 tenths. So 15 – 9 = 6.

As we can't take away 9 from 0, we borrow ten, so 10 – 9 = 1.

Example 2

	2	(2) 3	(1)0	• 8	6
-	1	0	7	• 5	4
£	1	2	3	• 3	2

6 – 4 = 2 and then 8 – 5 = 3

As we can't take away 7 from 0, we borrow ten.

So 10 – 7 = 3. Then 2 – 0 = 2 and 2 – 1 = 1.

Example 3

	(4) 5	(1)6	(6) 7	• (1)0	(1)2
-	1	7	4	• 2	9
£	3	9	2	• 7	3

As we can't take away 9 from 2, we need to borrow. As there are 0 tenths, we need to borrow 1 pound, which is ten 10ths. Borrowing one, we can do 12 – 9 = 3. There are 9 tenths now, so 9 – 2 = 7. Then 6 – 4 = 2. As we can't take away 7 from 6, we borrow one from the hundreds, so 16 – 7 = 9. Then 4 – 1 = 3

Example 4

	(4) 5	(1)6	(8) 9	• (1)3	4
-	3	8	7	• 5	1
£	1	8	1	• 8	3

4 – 1 = 3

As we can't take away 5 from 3, we borrow a pound. So 13 – 5 = 8. Then 8 – 7 = 1.

As we can't take away 8 from 6, we borrow one from the hundreds, so 16 – 8 = 8. Then 4 – 3 = 1

14 Fill the gaps.

403.87	112.90	186.40	530.80	376.46	497.35	365.54
- 339.08	- 78.90	- 176.54	- 74.65	- 233.73	- 156.65	- 363.45
463.80	374.66	105.36	436.44	583.72	378.26	473.64
- 374.51	- 46.27	- 0.93	- 6.57	- 37.46	- 312.98	- 437.37
478.47	450.47	476.44	362.46	374.55	159.80	365.48
- 375.83	- 437.83	- 400.00	- 75.36	- 47.46	- 47.69	- 357.67

15 These are the costs Emma has to pay for her conference:

Conference ticket: £135.45 **Travel:** £56.30

Calculate the total of these costs. Show your working.

____

16 Jack is comparing prices for the gym. He sees these monthly memberships:

Gym A	**Gym B**	**Gym C**	**Gym D**
£45.15	£50.50	£45.99	£51.75

Which gym has the lowest monthly membership? ____________________

17 An art class costs £79.99. There is a discount offer, "Sign up today and get £15 off." How much will the art class cost with this offer? Show your working.

____

18 A school has made money from a fundraising event. Here are the amounts raised:

Cake sale: £142.25 Raffle tickets: £98.00 School merchandise: £200.75

How much money has been raised in total from this event?

____

19 Martha has £75 to spend on groceries. She spends £41.30. How much money does Martha have left to spend?

____

20 Samira bought these items in a shop.

Pot	£18.85
Frying pan	£26.99
Mugs set of 4	£11.99

a.i) How much did these items cost altogether?

____

a.ii) Samira pays with a £50 and a £20 note. How much change does she get?

____

b) Jack is buying a shirt for £27.99 and a backpack for £19.99.

He pays with this note.

How much change will he get?

____

c) Zara is paying for food in a restaurant. She writes **17£**. What is wrong with her answer?

____

d) Dan is buying two shirts for £39.99 each. How much will he pay in total?

____

e) Chris thinks this is a £50 note. Is he correct?

____

f) Lou thinks she needs to pay 37p.

Is she correct?

____

g) Chloe bought a pen for £12 and a notebook for £24. She thinks she has spent 36p. Is she correct?

____

h) Gent thinks he has spent **£15p** while shopping. What is wrong with this?

____

21 a) Put 2p and 10p in the correct places on this number line.

b) Put £5 and £10 in the right places on this number line.

22 John orders food. It costs £12.95. He pays with a £20 note. How much change does he get?

____

23 Carly bought three drinks from the shop. A drink costs £1.69. How much did Carly pay for the drinks?

____

24 Your monthly budget for entertainment is £75.00. If you spend £48.25 in the first half of the month, how much can you spend in the second half.

____

Progress check

1 a) Add how many pounds and pence each of the values below is worth.

£4.50 is worth ____ pounds and ____ pence

£4.05 is worth ____ pounds and ____ pence

£0.45 is worth ____ pounds and ____ pence

b) Put the amounts above in order, from the lowest to the highest.

Lowest ________, ________, ________ Highest

2 You go to a shop and buy the two items shown below.

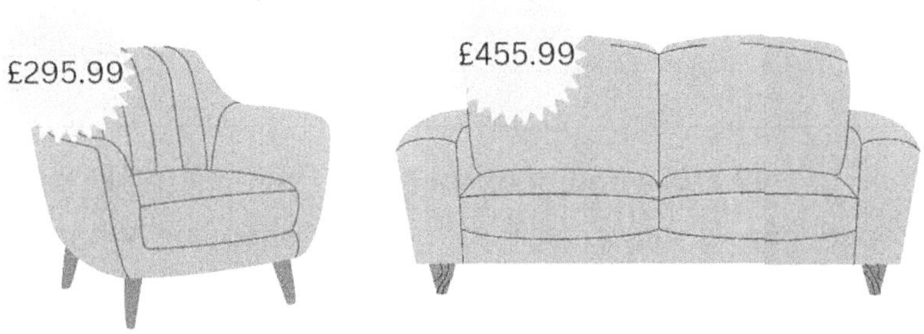

a) How do the sofa and the armchair cost in total?

____

b) You had a budget of £1000 for the two items. Is there any money left? If yes, how much?

____

3 Fill the gaps below. Include the unit.

£18.99 + £43.99 = ____	£66.83 – £34 = ____
£35.45 + ____ = £100	£79 – ____ = £69.70

4 Gurpreet lines up the cost of two different items to add them up. Here is what she has done:

```
  £56.99
+  £29.52
```

Is this correct? Explain your answer.

____

Round amounts of money to the nearest £1 or 10p

We round when:

- working out the total cost of items we have bought
- checking that we have done calculations correctly
- estimating the cost of bills over time
- estimating the time it takes to travel to a place

Round amounts of money to the nearest £1

To round to the nearest £1, means to decide which of the ones (units) the number is closest to.

Example: £34.30

£34.30 is between 34 and 35. As 34.30 is closer to 34, **to the nearest £1 that is £34**.

> To round an amount to the nearest pound, you decide by looking at the pence.
> If ending in 49p or less, you round the number **down**.
> If ending in 50p or more, you round the number **up**.
> Example: £1.49 is rounded to £1, but £1.50 is rounded to £2

Example: £67.88

£67.88 is between 67 and 68. As 67.88 is closer to 68, **to the nearest £1 that is £68**.

Example: £65.50

£65.50 is halfway between 65 and 66. £65.50 is the cut-off point for rounding up, so **to the nearest £1 that is £66**.

Example: £32.00

£32 is already a whole number. **To the nearest £1 that is £32**.

Example: £0.29

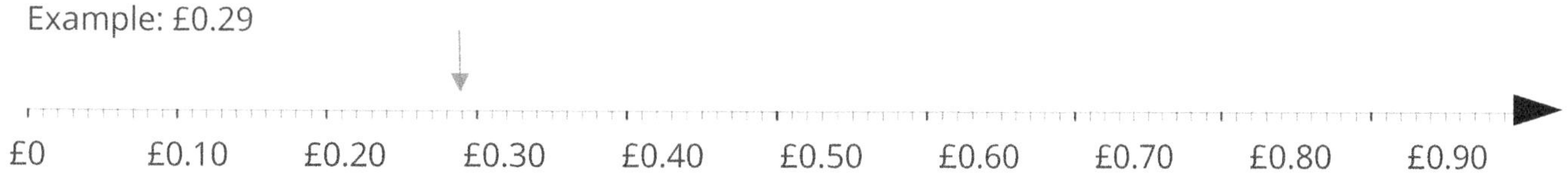

£0.29 is between 0 and 1. As 0.29 is closer to 0, **to the nearest £1 that is £0**.

Example: £40.49

£40.49 is between 40 and 41. As 40.40 is closer to 40, **to the nearest £1 that is £40**.

Example: £99.72

£99.72 is between 99 and 100. As 99.72 is closer to 100, **to the nearest £1 that is £100**.

Example: £248.29

£248.29 is between 248 and 249. As 248.29 is closer to 248, **to the nearest £1 that is £248**.

Example: £576.67

£576.67 is between 576 and 577. As 576.67 is closer to 577, **to the nearest £1 that is £577**.

1 Round these amounts to the nearest £1.

£83.57 ____	£19.03 ____	£4.36 ____	£27.26 ____	£35.50 ____	£50.49 ____
£76.00 ____	£54.65 ____	£90.69 ____	£99.80 ____	£0.36 ____	£0.75 ____
£63.50 ____	£70.00 ____	£24.10 ____	£168.56 ____	£376.84 ____	£527.12 ____

2 An amount has been rounded to the nearest pound. The rounded amount is £7. What could the original amount have been before it was rounded?

Lowest possible amount: ____ Highest possible amount: ____

3 An amount has been rounded to the nearest pound. The rounded amount is £23. What could the original amount have been before it was rounded?

Lowest possible amount: ____ Highest possible amount: ____

4 Leo buys a pack of stickers. Each pack costs £3.45. Round £3.45 to the nearest £1.

____

Rounding amounts of money to the nearest 10p

To round to the nearest 10p, means to decide which of the tenths the number is closest to.

Example: £34.31

£34.31 is between 34.30 and 34.40. As 34.31 is closer to 34.30 than to 34.40, **to the nearest 10p that is £34.30**.

> To round an amount to the nearest 10p, you decide by looking at the pence.
> If ending in 4p or less, you round the number **down**.
> If ending in 5p or more, you round the number **up**.
> Example: £1.42 is rounded to £1.40, but £1.45 is rounded to £1.50

Example: £67.88

£67.88 is between 67.80 and 67.90. As 67.88 is closer to 67.90, **to the nearest 10p that is £67.90**.

Example: £65.50

£65.50 is spot on £65.50 which is the cut off point for rounding up, so **to the nearest 10p that is £65.50**.

Example: £32.00

£32.00 is between 32.00 and 32.10. **To the nearest 10p that is £32.00**.

Example: £0.29

£0.29 is between 0.20 and 0.30. As 0.29 is closer to 0.30, **to the nearest 10p that is £0.30**.

Example: £40.49

£40.49 is between 40.40 and 40.50. As 40.49 is closer to 40.50, **to the nearest 10p that is £40.50**.

Example: £99.72

£99.72 is between 99.70 and 99.80. As 99.72 is closer to 99.70, **to the nearest 10p that is £99.70**.

Example: £248.29

£248.29 is between 248.20 and 248.30. As 248.29 is closer to 248.30, **to the nearest 10p that is £248.30**.

Example: £576.67

£576.67 is between 576.60 and 576.70. As 576.67 is closer to 576.70, **to the nearest 10p that is £576.70**.

5 Round these amounts to the nearest 10p.

£83.57 ____	£19.03 ____	£4.36 ____	£27.26 ____	£35.50 ____	£50.49 ____
£76.00 ____	£54.65 ____	£90.69 ____	£99.80 ____	£0.36 ____	£0.75 ____
£63.50 ____	£70.00 ____	£24.10 ____	£168.56 ____	£376.84 ____	£527.12 ____

6 Mina rents a car for the weekend. The cost per day is £37.85. What is £37.85 rounded to the nearest pound? ____________________

7 The cost of a bus ticket is £2.75. Round £2.75 to the nearest 10p. ____________________

8 James pays £76.99 for a pair of shoes. Round £76.99 to the nearest £1. ____________________

9 What is £789.49 rounded to the nearest pound? ____________________

10 Tanya is buying a shirt for £27.89, a coat for £43.39 and an umbrella for £5.99. Use rounding to estimate the total cost of all three items.

____

11 The price of a video game is £49.99. Round £49.99 to the nearest 10p. ______________________

12 Check these calculations by rounding to the nearest whole number. The first one has been done for you. You can use a calculator.

43.99 + 38.45 = 82.44	64.75 – 18.76 =	11.50 + 57.46 =	93.99 – 3.10 =
Checking:	Checking:	Checking:	Checking:
44 + 38 = 82	______________	______________	______________

13 Check these calculations by rounding to one decimal place. The first one has been done for you. You can use a calculator.

64.76 + 25.05 = 89.81	55.73 – 42.88 =	10.63 + 19.95 + 55.14 =	75.59 – 29.45 =
Checking:	Checking:	Checking:	Checking:
64.80 + 25.10 = 89.90	______________	______________	______________

14 An amount has been rounded to the nearest 10p. The rounded amount is £7.30. What could the original amount have been before it was rounded?

Lowest possible amount: _____ Highest possible amount: _____

15 An amount has been rounded to the nearest 10p. The rounded amount is £23.60. What could the original amount have been before it was rounded?

Lowest possible amount: _____ Highest possible amount: _____

16 Lisa earns £8.85 per hour. Estimate how much she earns from a nine-hour shift.

____

17 a) A garden needs a new hedge all around. The sides of the garden are measured as follows: 60m, 40m, 60m. What length of hedge will be needed?

____

b) The hedge is sold in 25m lengths. How many lengths will be needed?

____

c) Each length of hedge costs £29.99. Estimate the total cost.

____

18 A charity is raising funds through an event. They sold 90 tickets at £8.25 each. The organizer estimates they made £742.50. Is the organizer correct?

____

19 A shop sells apples for £0.89 per kg. John rounds this to the nearest 10p and says that the apples cost 90p per kg. Is John correct?

____

Progress check

1 Round these amounts.

£4.50 to the nearest pound is _____ .	£4.50 to the nearest 10p is _____ .
£4.05 to the nearest pound is _____.	£4.05 to the nearest 10p is _____.
£0.45 to the nearest pound is _____ .	£0.45 to the nearest 10p is _____ .

2 You go to a shop and buy the two items shown below.

a) How do the sofa and the armchair cost in total?

____

b) Check your answer using estimation.

____

3 a) Find the lowest and the highest amounts that could be rounded to the nearest £1 and give £6?

Lowest: _____ Highest: _____

b) Find the lowest and the highest amounts that could be rounded to the nearest 10p and give £6.80?

Lowest: _____ Highest: _____

4 Nicola rounded £73.99 to the nearest £1 and here is her answer: £70. Is this correct? Explain your answer.

____

Read time from analogue and 24-hour digital clocks in hours and minutes

We need to know how to read clocks so that we can plan our time.

This analogue clock shows it's 4:25.

The short hand tells us it is past 4 o'clock.

The long hand tells us it is 25 minutes **past** 4.

This analogue clock shows it's 4:50.

The short hand tells us it is past 4 o'clock.

The long hand tells us it is 10 minutes **to** 5.

There are 5 minutes in each main division of the clock.

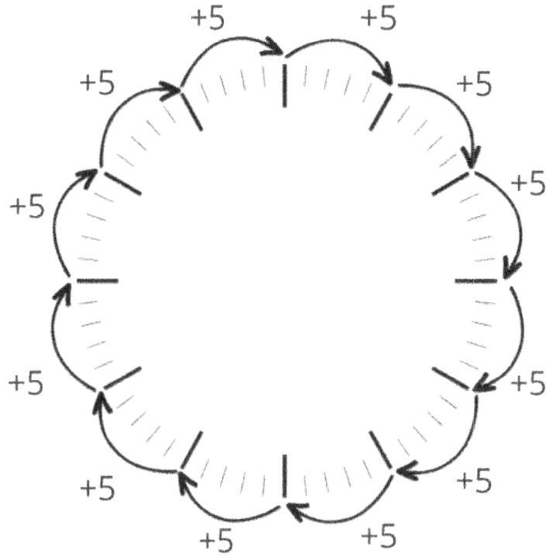

There are 60 minutes in one hour:
12 x 5 = 60

Notice how the hours hand has passed 4 but as it is not at 5 yet, that means it is still 4.

Time after half past can be read in more than way.

04:10	04:30	04:20	04:35	04:45
Four ten	Four thirty	Four twenty	Four thirty-five	Four forty-five
Ten past four	Half past four	Twenty past four	Thirty-five past four	Forty-five past four
			Twenty-five to five	Quarter to five

35 past 4 means 35 minutes **after** 4 o'clock.

25 to 5 means 25 minutes **to get to** 5 o'clock.

1 Match the clocks with the correct time.

03 : 05 07 : 40 06 : 20 03 : 55

2 Write the time under each clock.

________ ________ ________ ________

________ ________ ________ ________

3 Draw clock hands that show these times.

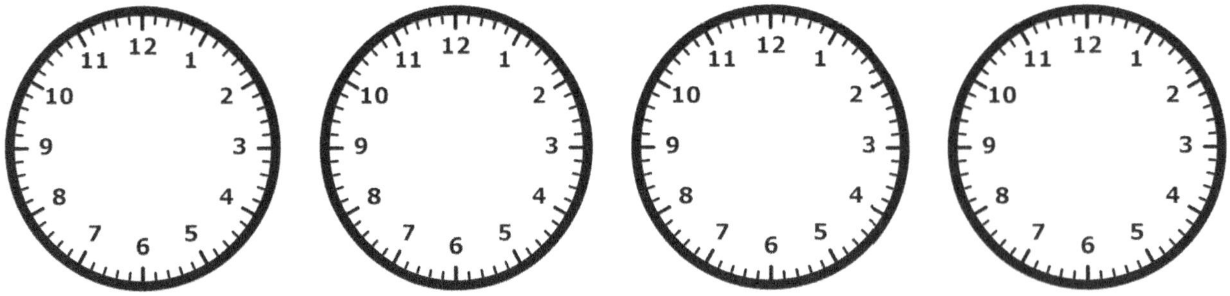

10 past 7 6 : 13 ten to 11 forty past 9

4 Match the clocks with the correct time.

21:12 | 19:45 | 14:30 | 07:12 | 08:55

quarter to eight | twelve past nine | 12 past 7 | fourteen thirty | 5 to 9

5 Match the clocks with the correct time.

17:33 | 08:19 | 04:21 | 01:16

6 Write the time under each clock, in two different ways.

________ ________ ________ ________

________ ________ ________ ________

08:35 | 10:06 | 20:42 | 16:50

________ ________ ________ ________

________ ________ ________ ________

7 Draw the clocks that show these times.

: | : | : | :

12 past 7 | fourteen thirty | 5 to 9 | 8 past 7

8 a) A GP opens in the morning at the time shown on this clock. What time does the GP open?

____

b) This clock shows the time for the evening swim session at the local pool. What time does the clock show?

__________pm

c) Emma and her daughter arrive at the zoo at a quarter past three in the afternoon. Which clock shows this time? Tick it.

d) Mary arrives at the shopping centre at 10:35. She plans to spend 1 hour 45 minutes shopping. Will she finish shopping by 12:00? Explain your answer.

____

____

e) George looks at his watch before heading to work. What time is it?

________pm

f) A client called Mrs. Green phoned to make an appointment. This clock shows the time she phoned.

Mrs. Green says she will come to the office in 3 hours. Write a note to say when Mrs. Green will come to the office. ______________________

g) Four friends played a video game and finished with the following times:

Friend A - 2 hours 40 minutes
Friend B - 3 hours
Friend C - 2 hours 45 minutes
Friend D - 2 hours 50 minutes

Who had the shortest time?

____

h) A conference is scheduled to start at 09:30. This is the schedule for it:

Morning session: 2 hours 30 minutes
Lunch break: 1 hour
Afternoon session: 2 hours 30 minutes

Calculate what time the lunch break will start.

____

i) Peter arrives at the football stadium at 2:15 PM.

The game is expected to last 1 hour and 30 minutes.

Will he be able to catch his 4:00 PM train? Explain your answer.

____

j) A customer named Mrs. Smith phoned to reserve a table for dinner. This clock shows the time she phoned. Mrs. Smith says she will come to the restaurant in 1 hour and 45 minutes. Write a note to say when Mrs. Smith will come to the restaurant. Show your working out.

____

9 Sally thinks this clock shows 2:20 in the afternoon. Is she correct? Explain.

____

10 Michael gets to the gym at 7:45 AM. He plans to do a 1 hour and 30 minutes workout. Michael wants to finish his workout by 9:00 AM. Does he have enough time to finish his workout? Give a reason for your answer.

____

11 It's 3:35 PM. Jack thinks he needs to wait 65 minutes until it's 4:00 PM. Is Jack correct? Explain.

____

Progress check

1 Write underneath each clock face the time shown, in 12-hour format (with AM/PM).

__________ __________ __________ __________

2 Write underneath each clock face the time shown, in 24-hour format.

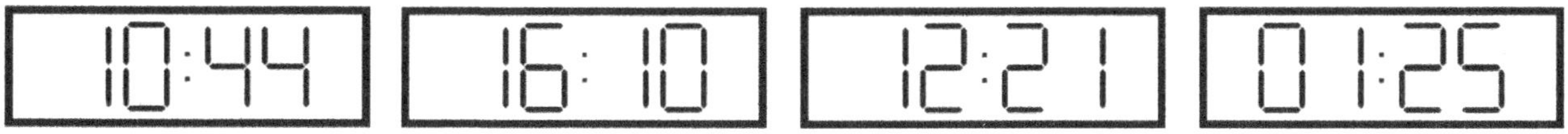

__________ __________ __________ __________

3 Write 'quarter to eight' in another two ways, in words (not digits).

__________, __________

4 a) Show 'twenty-five past 9' in this clock face.

b) Zen thinks that this clock face below shows the time '5 to 9'.

Is he correct? Why?

____

____

Read, measure and record time using am and pm

We have two formats for reading time. The 24 hour format and the 12 hour format.

The 24-hour format goes up to 24:00. As there are 24 hours in a day, this format starts at midnight and goes back full circle at midnight.

24-hour format		12-hour format
00 : 00	midnight	12 am
01 : 00		1 am
02 : 00		2 am
03 : 00		3 am
04 : 00		4 am
05 : 00		5 am
06 : 00		6 am
07 : 00		7 am
08 : 00		8 am
09 : 00		9 am
10 : 00		10 am
11 : 00		11 am
12 : 00	midday	12 pm
13 : 00		1 pm
14 : 00		2 pm
15 : 00		3 pm
16 : 00		4 pm
17 : 00		5 pm
18 : 00		6 pm
19 : 00		7 pm
20 : 00		8 pm
21 : 00		9 pm
22 : 00		10 pm
23 : 00		11 pm
24 : 00		12 pm

AM means 'ante meridiem' in Latin, which means before noon or before midday.

PM means 'post meridiem' in Latin, which means after noon or after midday.

The 12-hour format goes up to 12:00. From midnight to midday there are 12 hours and another 12 hours from midday to midnight.

These clocks show 8 o'clock in the morning.

These clocks show 8 o'clock in the evening.

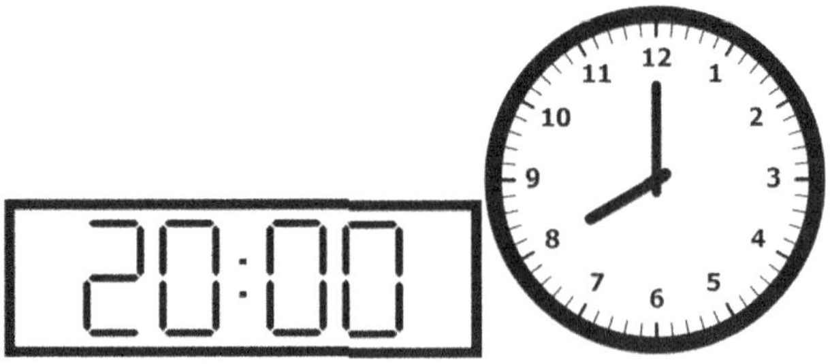

1 Write these morning times in the 12-hour format.

05:30	07:42	09:13	10:48	11:06
______	______	______	______	______

2 Write these afternoon/evening times in the 12-hour format.

13:09	18:13	15:26	20:59	23:35
______	______	______	______	______

3 Write these morning times in 24-hour and 12-hour formats.

______ ______ ______ ______

______ ______ ______ ______

4 Write these afternoon/evening times in 24-hour and 12-hour formats.

______ ______ ______ ______

______ ______ ______ ______

5 Chris sets off for a road trip at 9:00 AM. The trip takes 3 hours and 45 minutes. What time will he arrive at his destination? Use the 12-hour clock format.

____

6 A football match begins at 3:30 PM and ends at 5:00 PM, how long does the match last?

____

7 Matt starts watching a 2-hour movie at 7:45 PM. What time will the movie end?

---

8 Emma's flight departs at 11:55 PM. She needs to be at the airport 2 hours before departure. It takes her 45 minutes to reach the airport. What is the latest time she should leave her house?

---

9 John wakes up at 6:30 AM. He takes 45 minutes to get ready and then walks to the bus stop. The bus arrives 12 minutes after he reaches the bus stop. What time does the bus arrive?

---

10 A bakery opens at 7:30 AM and closes at 5:00 PM. How long is the bakery open each day?

---

11 If a night shift at a factory starts at 10:30 PM and ends at 7:00 AM, how long does the shift last?

---

12 Bill needs to be at the airport by 7:00 PM for a 9:00 PM flight. It takes him 1 hour to drive to the airport and he needs to find a parking spot, which takes about 15 minutes. He believes he can leave his house at 6:00 PM and arrive on time. Is he correct?

---

13 Mike has to be at a meeting by 3:15 PM. If it takes him 35 minutes to drive there, what is the latest time he can leave to make it to his meeting on time?

---

14 The football game starts at 8:15 PM and it lasts for 1 hour and 45 minutes. What time will the game end?

---

15 The train leaves the station at 5:20 AM. The journey takes 3 hours and 15 minutes. What time does the train arrive at the destination?

---

16 A store opens at 8:30 AM. It closes 9 hours later. What time does the store close?

---

17 Tanya arrives at the office at 8:45 AM. She leaves work at 5:00 PM. If she takes an hour for lunch, how long does she work each day?

Progress check

1 Write underneath each clock face the time shown, in 24-hour format.

2 a) You have an interview at 12:30pm. It is now 11:12am. How long do you have until the start of the interview?

b) At the interview, you will plan for 20 minutes a presentation, which you will present for 10 minutes and then answer questions for 30 minutes. How long will the interview take in total?

3 a) What time will it be 30 minutes after 9:10pm?

b) What time will it be 2 hours before 10:45am?

4 a) Enes is making pancakes. It takes 3 minutes to make one pancake. How long would it take Enes to make 12 pancakes?

b) Enes has 1 hour available for making pancakes. How many pancakes can he make in one hour?

5 Enes needs to finish making the pancakes by 11:00. What time does he need to start making the pancakes, if it takes him 1 hour to make the pancakes and 10 minutes to make the mixture? He starts making the pancakes at 09:55. Is he on time?

Compare metric measures of length including millimetres, centimetres, metres and kilometres

We use millimetres, centimetres, metres and kilometres to measure:

- the length or width of furniture
- the distance between places
- the length of fabric
- the height of people

We use these instruments to measure length:

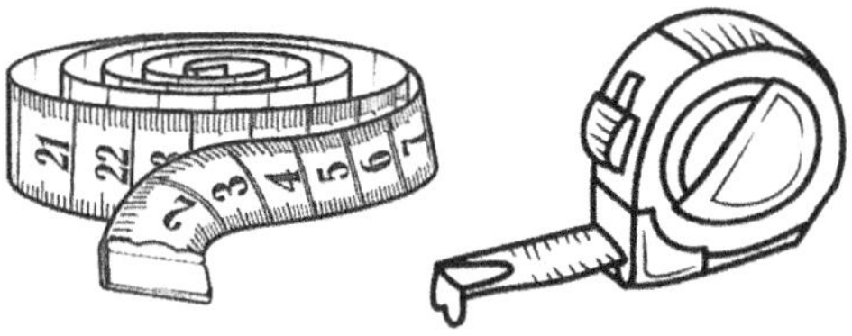

measuring tape or tape measure

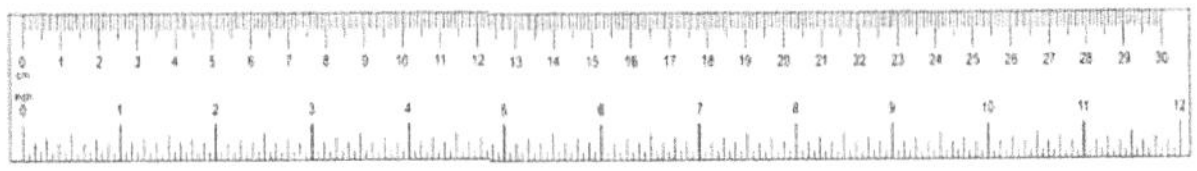

ruler

To measure, you align the ruler or measuring tape with the item you are measuring.

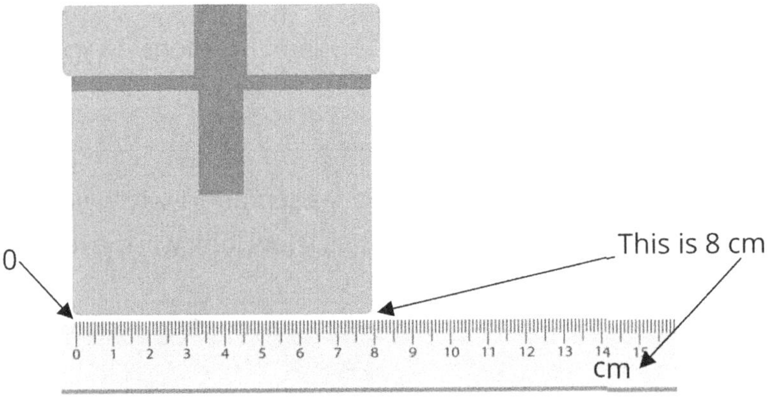

1 Measure these lines using a ruler.

___ cm

___ cm

___ cm

___ cm

___ cm

___ cm

___ cm

In a ruler, you will usually see millimetres (mm) and centimetres (cm). This ruler measures up to 30cm.

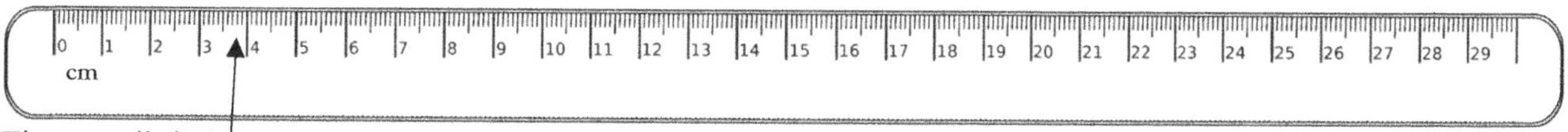

The small divisions represent millimeters.
Can you count how many millimetres there are in one centimetre?

There are 10mm in 1cm.
The width of a nail is about 1cm

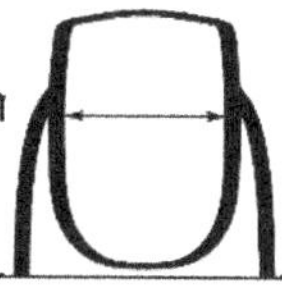

The depth of a fingernail is about 1mm.

You can use a ruler to help you remember this fact. You can even have a ruler in the exam room.

To convert 5cm into mm:

This is equal to 50mm. 5 x 10 = 50

To convert 80mm into cm:

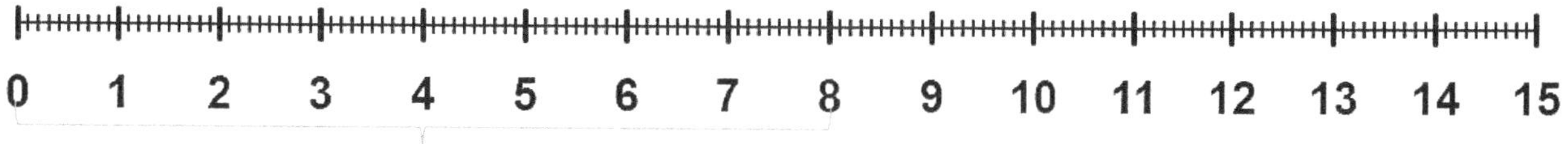

This is equal to 8cm. 80 ÷10 = 8

2 Jeta is posting a package. The maximum length the box can have is 18cm. She wants to use the longest box for posting.

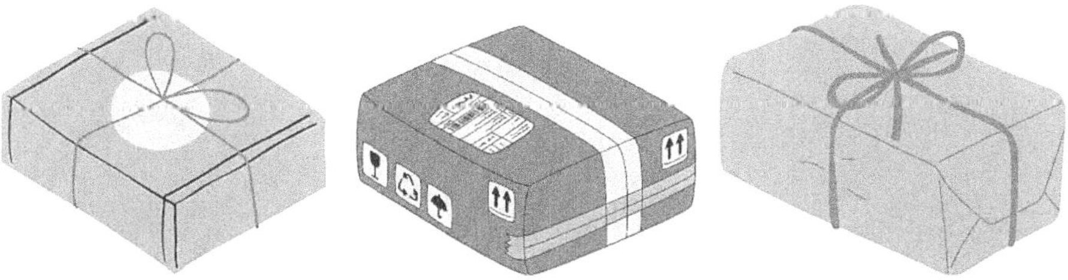

15.5cm 190mm 16.2cm

Which of the boxes can Jeta post?

---

3 Zain needs a 60mm wire. These are the options:

5.5cm 70mm 4.5cm 40mm

Which of these should Zain buy? Why?

---

4 How wide is this picture frame? Don't forget to add the unit in your answer.

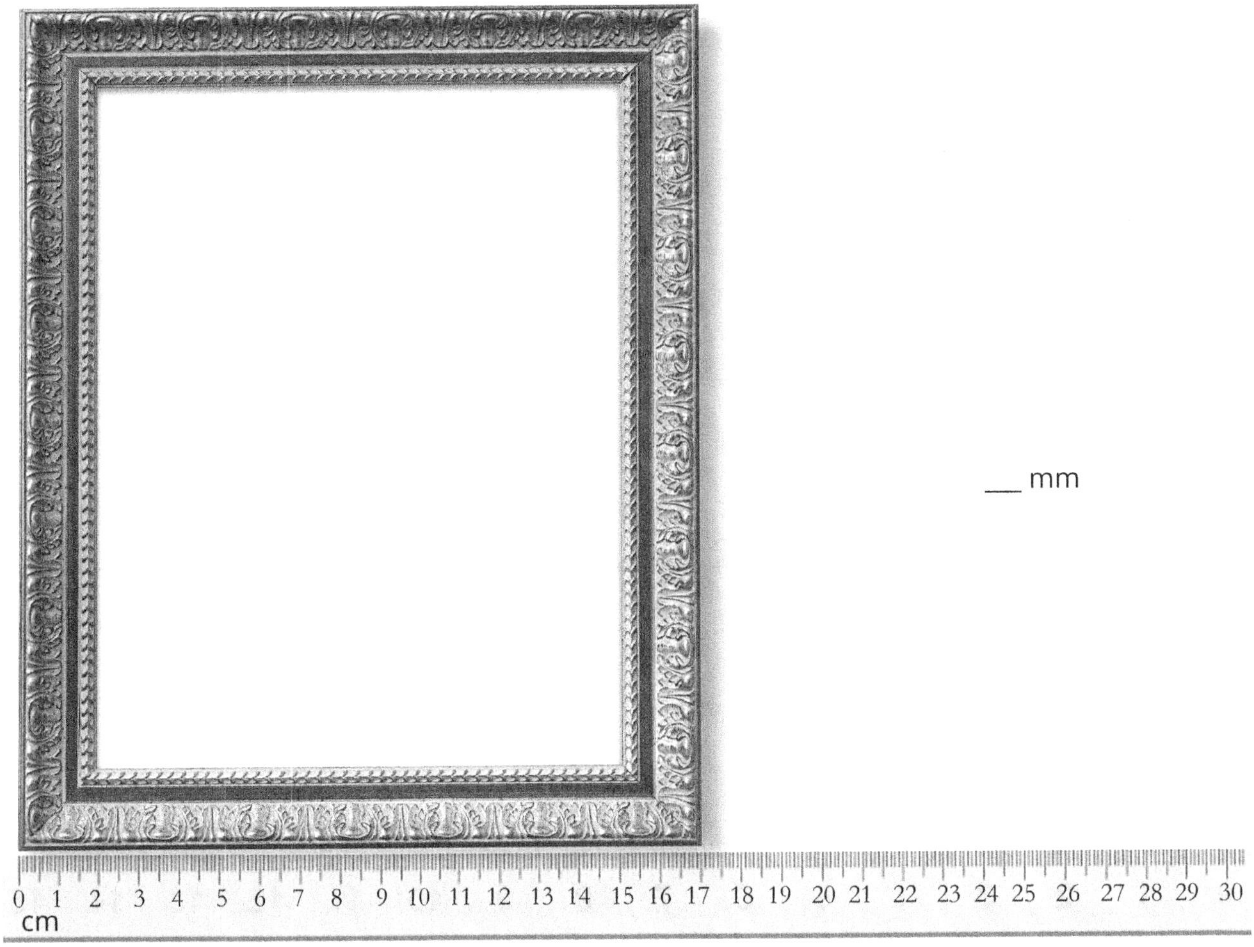

___ mm

5 Which of these is the longest? Circle your answer.

487m   485m   472m   549m

6 Put these suitcases in order of width, starting with the narrowest.

620mm   59cm   602mm   63.1cm

__________, __________, __________, __________

7 Circle the shortest measurement:

0.15cm   0.10cm   0.1cm   1.0cm

To measure the distance between places, we use kilometres (km). A measuring tape is very similar to a ruler, only longer. You will see cm, mm and m in a measuring tape.

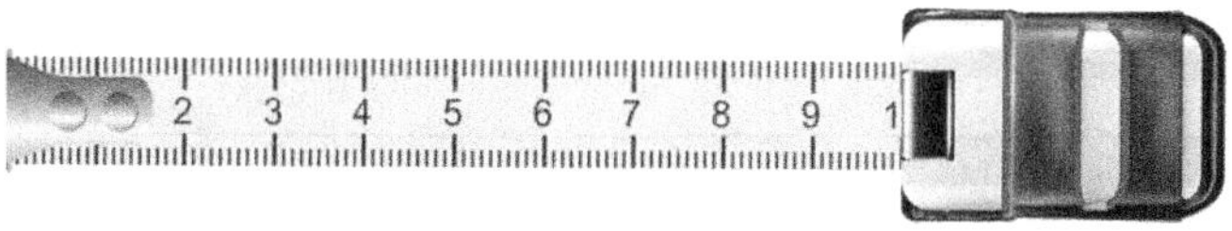

The small divisions represent millimetres. The bigger divisions

100cm = 1m
Centi, like **cent** or **century**, means 100.
That's why 100cm in 1m.
The span between shoulder and the fingers on the other hand, is about 1 metre.

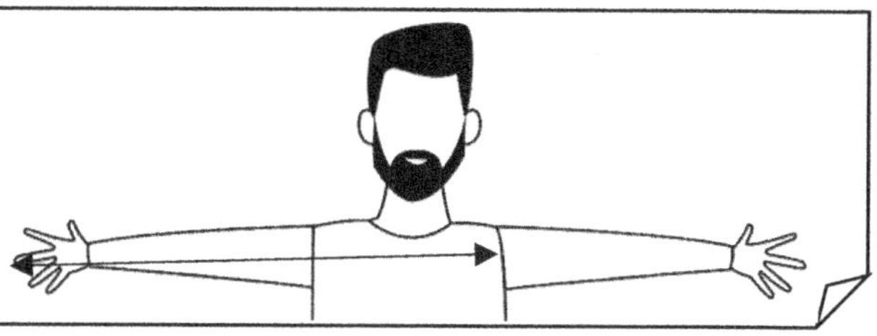

To convert cm into m, you **divide the measurement in cm by 100**, or follow this method:

500cm into m	Start by writing the conversion fact.	**650cm into m**	Start by writing the conversion fact.
1m = 100cm (x5) __m = 500cm	Under 100cm, write 500cm. 100 **x 5** = 500cm, so 1 **x 5** = 5m	1m = 100cm (x6.5) __m = 650cm	Under 100cm, write 650cm. 100 **x 6.5** = 650cm, so 1 **x 6.5** = 6.5m

To convert cm into m, you **multiply the measurement in m by 100**, or follow this method:

8**m into cm**	Start by writing the conversion fact.	**2.3m into cm**	Start by writing the conversion fact.
(x8) 1m = 100cm 8m = ___cm	Under 1m, write 8m. 1 **x 8** = 8m, so 100 **x 8** = 800cm	(x2.3) 1m = 100cm 2.3m = ___cm	Under 1m, write 2.3m. 1 **x 2.3** = 2.3m, so 100 **x 2.3** = 230cm

8 Fill the gaps.

0.1 m = 10 cm	0.2 m = cm	0.3 m = ___cm	0.4 m = ___cm	0.5 m = ___cm
0.6 m = ___cm	0.7 m = ___cm	0.8 m = ___cm	0.9 m = ___cm	0.11 m = ___cm
0.25 m = ___cm	0.80 m = ___cm	___m = 65 cm	___m = 82 cm	___m = 641cm
___m = 947 cm	___m = 163 cm	___m = 394 cm	___m = 370 cm	___m = 468 cm

9 Which of these is the longest? Tick it.

87cm	91cm	130cm	1.5m

10 Atif is buying a new wardrobe. The space where the wardrobe will go is 156cm wide. These are the widths of the wardrobes Atif can afford.

A	B	C	D
1.5m	160cm	1.6m	1.55m

Tick the wardrobes that would fit into the space.

11 Jen is putting two identical bookcases next to each other in the living room. Each bookcase is 86.2 cm wide. The space Jen has available for the two bookcases is 1.8m. Is there enough space to put the two bookcases?

____

12 Zain needs a new storage cabinet. The space where he will put the cabinet is 85cm long. These are the cabinets that Zain likes.

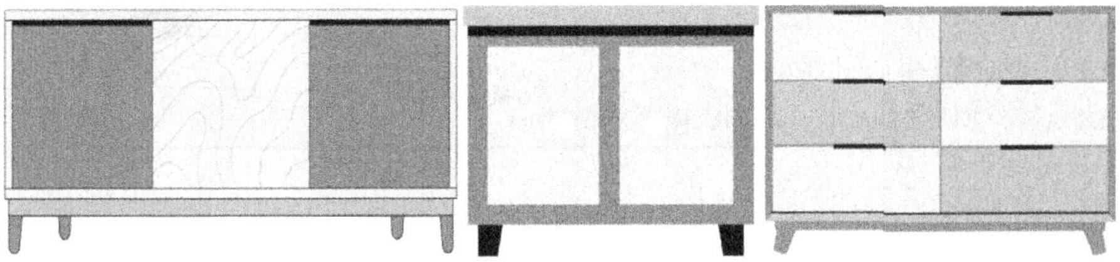

0.83m　　68cm　　800mm

He wants to buy the longest cabinet that fits the space. Tick the cabinet Zain should buy.

13 Circle the shortest measurement:

0.98m　　114cm　　0.95m　　1.15m

14 John has 4 planks of wood in different lengths:

a) 2.2 meters　　b) 170 centimeters　　c) 2.5 meters　　d) 191 centimeters

Which plank is closest to 2 meters long? __________________________________

15 Gela is shopping for a lamp shade. These are the ones she likes. Gela wants to buy the shortest one.

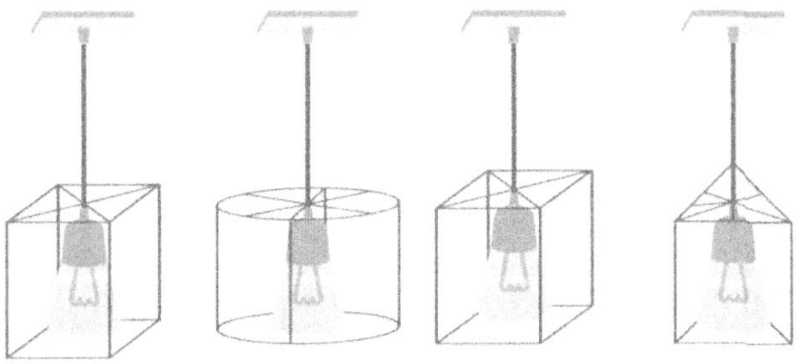

1.51m　　185cm　　2.0m　　168cm

16 A swimming pool has a depth of 1.25 meters at the shallow end and 2.04 meters at the deep end. How much deeper is the deep end compared to the shallow end?

____

17 Jenny wants to install a shelf and a lamp in her room. The width of the room is 2.4 m. The width of the shelf is 1.5 m and the lamp requires 80 cm of space. Will both the shelf and the lamp fit in the room? Show your working.

____

Kilometres are much longer than metres.
We use kilometres to measure long distances.

1km = 1000m

To convert m into km, you **divide the measurement in m by 1000**, or follow this method:

5000m into km	Start by writing the conversion fact.	**6500m into km**	Start by writing the conversion fact.
1km = 1000m __m = 5000m (x5)	Under 1000m, write 5000m. 1000 **x 5** = 5000m, so 1 **x 5** = 5km	1km = 1000m __km = 6500m (x6.5)	Under 1000m, write 6500m. 1000 **x 6.5** = 6500m, so 1 **x 6.5** = 6.5km

To convert km into m, you **multiply the measurement in m by 1000**, or follow this method:

8km into m	Start by writing the conversion fact.	**2.3km into m**	Start by writing the conversion fact.
(x8) 1km = 1000m 8km = ___m	Under 1m, write 8km. 1 **x 8** = 8km, so 1000 **x 8** = 8000m	(x2.3) 1km = 1000m 2.3km = ___m	Under 1m, write 2.3m. 1 **x 2.3** = 2.3km, so 1000 **x 2.3** = 2300m

18 Fill the gaps.

0.1km = 100m	0.2km = ___m	0.3km = ___m	0.4km = ___m	0.5km = ___m
0.6km = ___m	0.7km = ___m	0.8km = ___m	0.9km = ___m	0.11km = ___m
0.25km = ___m	0.80km = ___m	___km = 650m	___km = 820m	___km = 641m
___km = 947m	___km = 163m	___km = 3940m	___km = 3705m	___km = 4689m

19 Lora is planning to visit four different towns in the next few weeks. She will visit them in order of distance, starting with the shortest.

Town A	**Town B**	**Town C**	**Town D**
73.5km	70.2km	68.9km	74.0km

Put the towns in the order that Lora will visit them. ________, ________, ________, ________

20 Lisa walks from her house to the library, which is 1.3 kilometers away. Then she walks to the grocery store, which is another 0.7 kilometers away. How far does she walk in total?

____

21 Cala works in different offices each day. These are the distances of each office from her home.

6.2km   8.9km   10.1km   7.5km

Put these distances in order, starting with the shortest: _______, _______, _______, _______

22 Write the short form of each unit.

kilometre	millimetre	metre	centimetre
_____	_____	_____	_____

23 Put these units in order, starting with the shortest.

kilometre millimetre metre centimetre

__________, __________, __________, __________

24 Noor lives 3.7km away from her sister. She is visiting her sister today and returning tomorrow. What is the total distance Noor is travelling? Include the unit in your answer.

____

25 Tick the correct option.

a) The height of a tree is about:

3mm 3cm 3m 3km

b) The length of an earring is about:

2mm 2cm 2m 2km

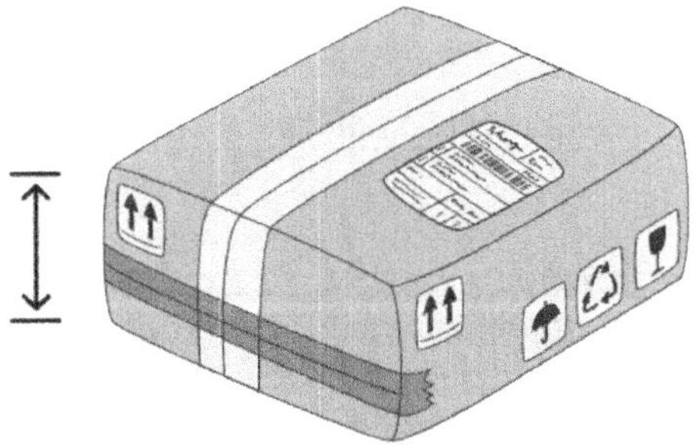

c) The depth of this package that fits in the letterbox is about:

20mm 20cm 20m 20km

d) The height of a toddler is about:

80mm 80cm 80m 80km

26 Zoi works in different offices from Monday to Thursday. These are the distances from her home to each office.

A	B	C	D
1.5km	1.65km	952cm	1.95m

Which of the offices is furthest away from her home? ______________________

27 Jacob needs to put a shelf on his wall. The wall space is 200 cm wide. He has three shelves to choose from:

Shelf A: 2.1 meters	Shelf B: 210 cm	Shelf C: 1.9 meters

Jacob thinks Shelf A will fit perfectly on his wall because 2.1 is less than 200. Is Jacob correct? Explain your answer.

____

Progress check

1 Fill in the gaps below.

____ m = 600 cm   450 cm = ____ m   ____ km = 2400 m

2 Ayan and Ayla are measuring their heights.

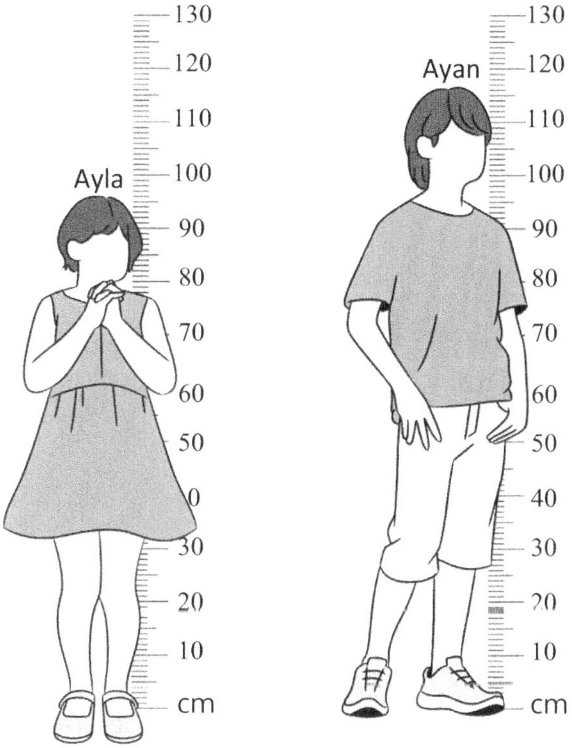

a) Write Ayan's height in centimetres. ______________________________

b) What is Ayan's height in metres? ______________________________

c) What is the difference between Alyla's and Ayan's heights? ______________________________

3 Put these measurements in order, from the smallest to the largest: 64cm, 0.8 m, 700 mm .

__________, __________, __________

4 Ali measured the space where she can fit a wardrobe in her bedroom and it was 85cm. She finds a wardrobe that is 87cm long and she says: 'I can buy this wardrobe as it would fit in my bedroom and there would be 2 cm left'. Is this correct? Explain your answer.

____

Compare measures of weight including grams and kilograms

We measure the weight of:

- ingredients when cooking
- parcels when posting
- groceries when shopping
- luggage

We use these instruments to measure weight:

kitchen scale

bathroom scale

luggage scale

postal scale

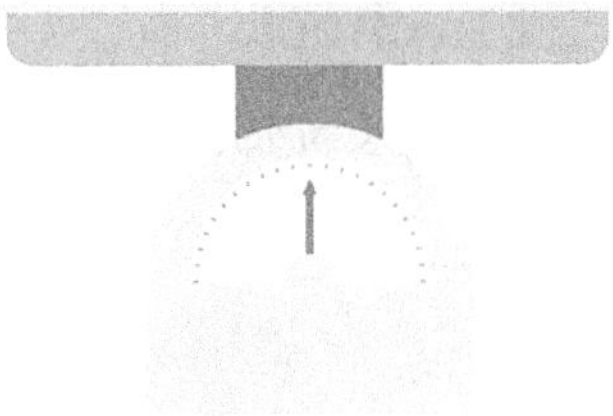

baby scale

When measuring weight, you look at:

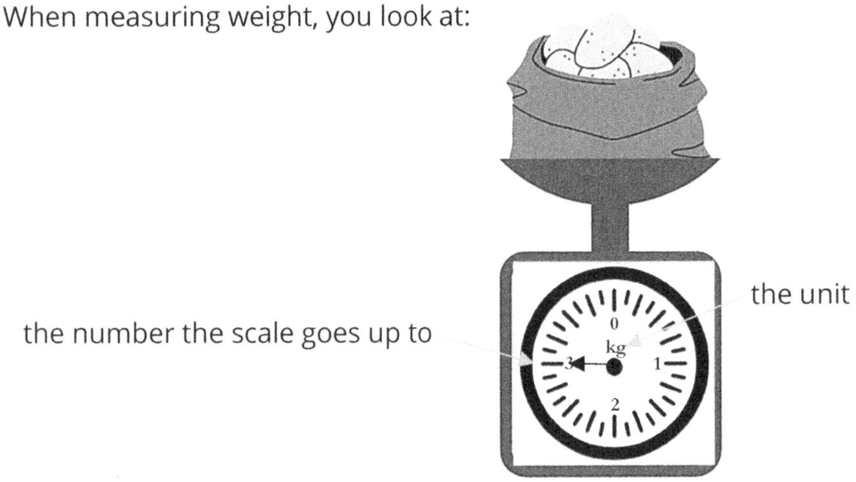

This bag of potatoes weighs 3 kilograms.

Some scales are digital.

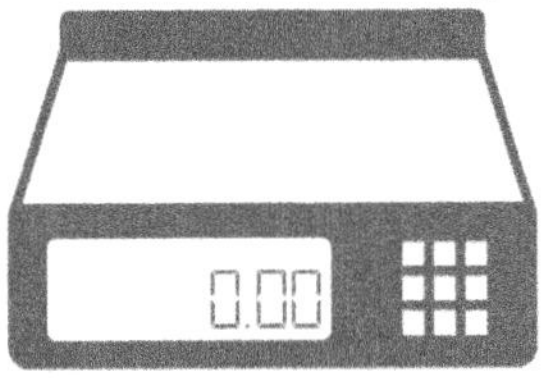

Some scales are analogue, just like clocks.

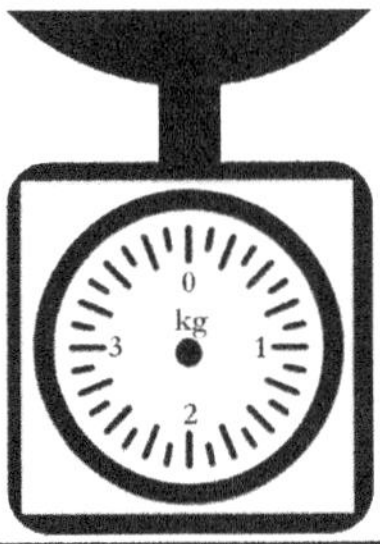

We use two main units to measure weight: grams (g) and kilograms (kg).

1 **gram** weighs less than 1 **kilogram**. There are one thousand grams in 1 kilogram.

Two paracetamol tablets weigh 1grar A bag of sugar usually weighs 1 kilogram

1 Write the short form of each unit.

kilogram ____ gram ____

2 Which unit is heavier: 1 kilogram or 1 gram? ____________________

3 Put these units in order, starting with the heaviest.

5 kilograms 5 grams 50 grams 50 kilograms

________, ________, ________, ________

To convert g into kg, you **divide the measurement in g by 1000**, or follow this method:

5000g into kg	Start by writing the conversion fact.	**6500g into kg**	Start by writing the conversion fact.
1kg = 1000g ↓x5 __g = 5000g	Start by writing the conversion fact. Under 1000g, write 5000g. 1000 **x 5** = 5000g, so 1 **x 5** = 5kg	1kg = 1000g ↓x6.5 __kg = 6500g	Start by writing the conversion fact. Under 1000g, write 6500g. 1000 **x 6.5** = 6500g, so 1 **x 6.5** = 6.5kg

To convert kg into g, you **multiply the measurement in m by 1000**, or follow this method:

8kg into g	Start by writing the conversion fact.	**2.3kg into g**	Start by writing the conversion fact.
x8↓ 1kg = 1000g 8kg = __g	Start by writing the conversion fact. Under 1kg, write 8kg. 1 **x 8** = 8kg, so 1000 **x 8** = 8000g	x2.3↓ 1kg = 1000g 2.3kg = __g	Start by writing the conversion fact. Under 1kg, write 2.3kg. 1 **x 2.3** = 2.3kg, so 1000 **x 2.3** = 2300g

4 Fill the gaps.

0.1kg = 100g	0.2kg = __g	0.3kg = __g	0.4kg = __g	0.5kg = __g
0.6kg = __g	0.7kg = __g	0.8kg = __g	0.9kg = __g	0.11kg = __g
0.25kg = __g	0.80kg = __g	__kg = 650g	__kg = 820g	__kg = 641g
__kg = 947g	__kg = 163g	__kg = 3940g	__kg = 3705g	__kg = 4689g

5 Read the measurements in these scales.

__kg __g　__kg __g　__kg __g　__kg __g

6 Put the weights of these suitcases in order, starting with the lightest.

22.5kg 19.8kg 21.0kg 20.9kg

________, ________, ________, ________

7 Read the measurements in these scales. Don't forget to add the unit.

________________ ________________ ________________

8 Jeta is posting a package. The maximum weight the boxes can have is 2kg.

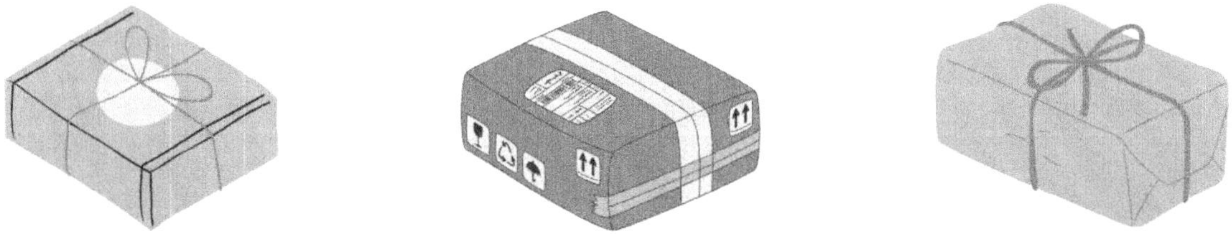

1.5kg 1900g 2.2kg

Which of the boxes can Jeta post?

____

9 Gary is making cakes. He needs 1kg of sugar. He has 650g of sugar. How much more sugar does he need?

____

10 Kelly is posting a pallet with furniture to another city. The pallet weighs 17.8kg. The furniture weighs 82.5kg. Kelly thinks that the pallet and the furniture weigh less than 100kg altogether. Is Kelly correct?

____

11 Circle the correct answer.

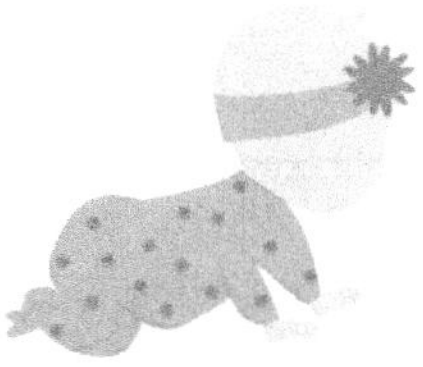

a) The weight of a baby is about:

10g 10kg

b) The weight of a plate of pasta is about:

100g 100kg

c) The weight of a feather is about:

1g 1kg

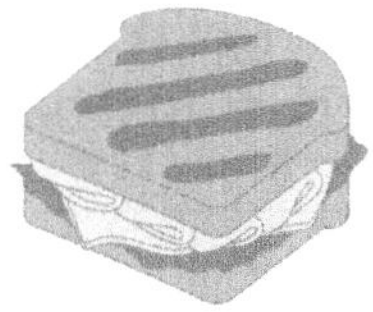

d) The weight of a sandwich is about:

200g 200kg

12 Sara thinks a TV unit weighs 20g. What is wrong with her answer?

____

13 Charles thinks 10.0kg is the same as 10kg. Is he correct?

____

14 Paul is sending a letter by post. The maximum weight should be 100g. He is sending 3 cards, which weigh 30g each. The weight of the envelope is 7g. Can Paul send the letter?

____

15 Tracey thinks this scale shows 65kg. Is Tracey correct? Why?

____

____

16 A gym trainer can lift weights up to 50 kg. Which of these weights can he lift? Tick the correct ones.

30 kg 3200 g 51000 g 4800 g

17 Annie is at a supermarket buying jam. The store sells jam jars weighing 200 g, 450 g, 800 g, and 1 kg. Annie wants the smallest jar that is over 500 g, which jar should she buy?

____

18 Daisy goes to the vet every month for a check-up. After four visits, her weights recorded were: 5.2 kg, 5.22 kg, 5.3 kg, and 5.19 kg. Which is the lightest weight?

____

19 A package of rice weighs 1.5 kg. Jimmy says, "This package of rice weighs 1500 g, so it is heavier than a 1 kg bag of flour because 1500 is a larger number than 1." Is Jimmy correct? Explain your answer.

____

20 Sarah weighs two boxes. Box A weighs 2500 grams and Box B weighs 2.4 kilograms. Sarah thinks that Box A is lighter than Box B because 2.4 is a bigger number than 2.5. Is Sarah correct? Give a reason for your answer.

____

21 A bag of potatoes weighs 4 kg. Tom adds 300 grams of onions to the bag. He then says, "The bag now weighs 4.3 kg because I have added 0.3 kg to it." Is Tom correct? Explain your answer.

____

Progress check

1 Fill in the gaps below.

4kg = ___ g	3500g = ___ kg

2 The weight of some pears and apples is shown in the scales below.

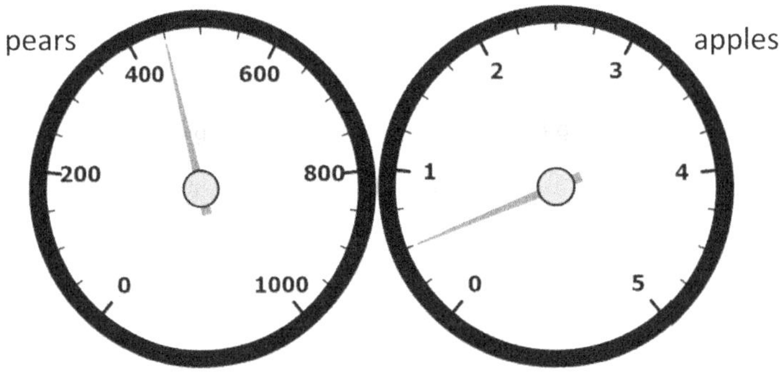

a) What is the pears' weight in kilograms? ______________

b) What is the pears' weight in grams? ______________

c) What is the difference between the pears and apples' weight? ______________

3 Put these measurements in order, from the smallest to the largest: 6.4kg, 400g, 6500g, 0.9kg .

_______, _______, _______, _______

4 Aaron thinks that these potatoes weigh 14kg.

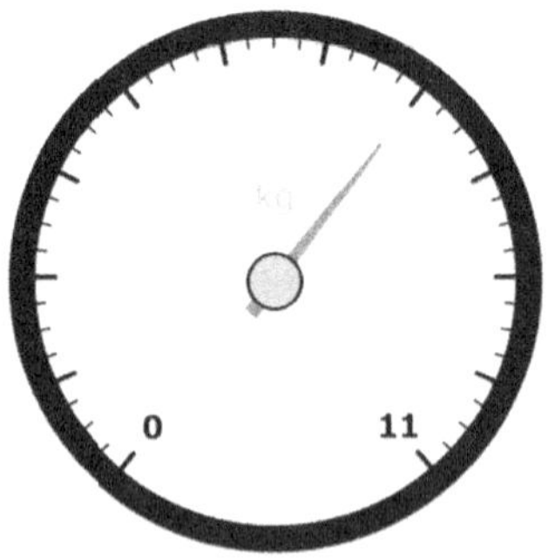

Is he right? Give a reason for your answer.

____

Compare measures of capacity including millilitres and litres

We measure the capacity of items when:

- cooking
- storing liquid
- using bottles
- getting petrol or diesel
- administering medicine
- gardening
- painting walls

We use these instruments to measure capacity.

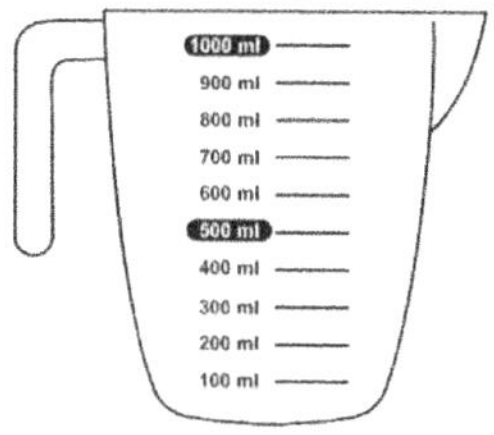

measuring jug

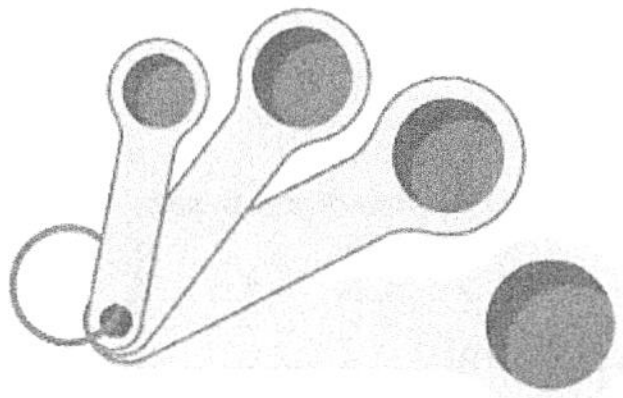

measuring spoons

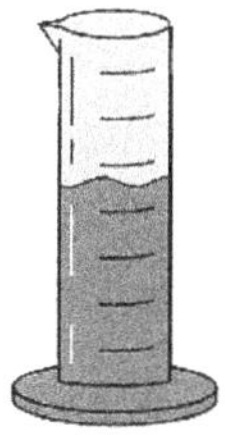

cylinder

syringe

To measure capacity:

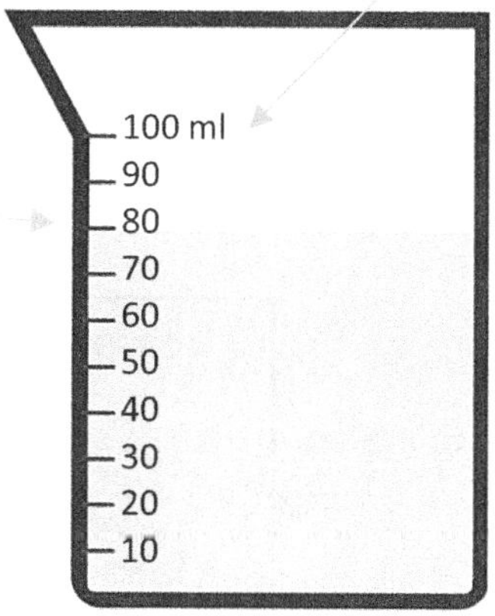

This jug holds 80ml of liquid.

We use two main units to measure capacity: milliliters (ml) and litres (l)

1 **millilitre** holds less than 1 **litre**.

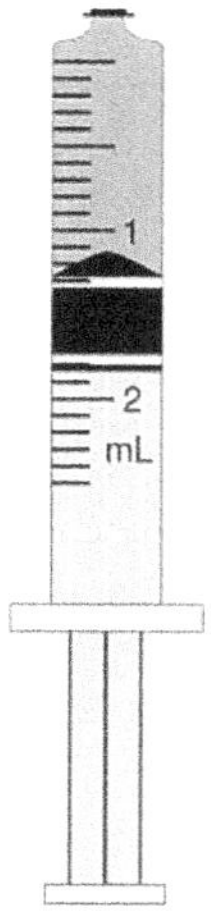

Medicine is measured in millimetres.

A bottle of oil usually holds 1 litre.
A litre is greater than a millilitre.
There are one thousand millilitres

1. Write the short form of the units.

   litre ____ millilitre ____

2. Which is greater: 1 litre or 1 millilitre? ____________

3. Put these units in order, starting with the heaviest.

   5 litres 5 millilitres 50 millilitres 50 litres

   ________, ________, ________, ________

4 How much liquid is there in each container? Include the unit.

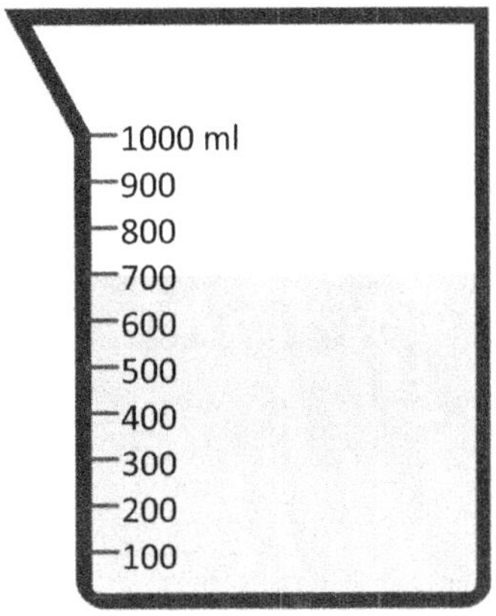

____

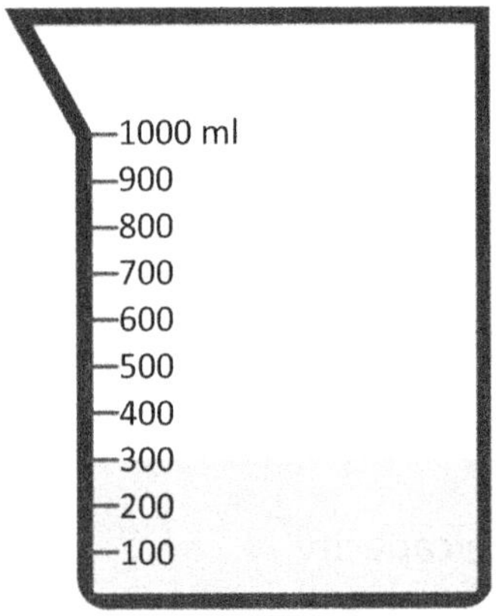

____

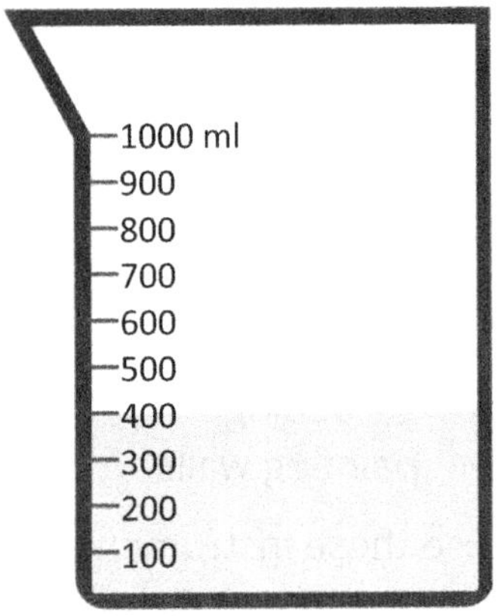

____

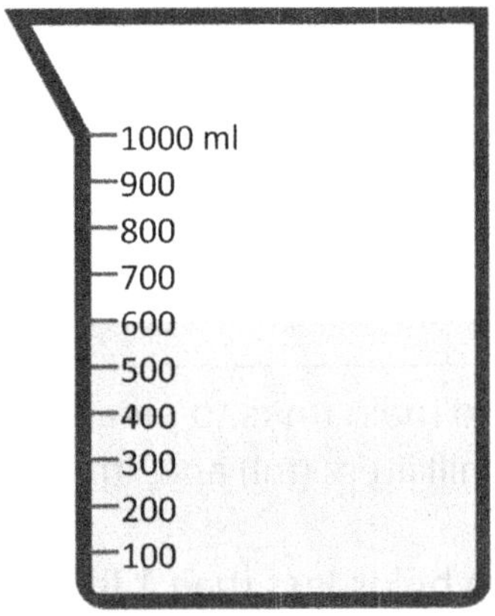

____

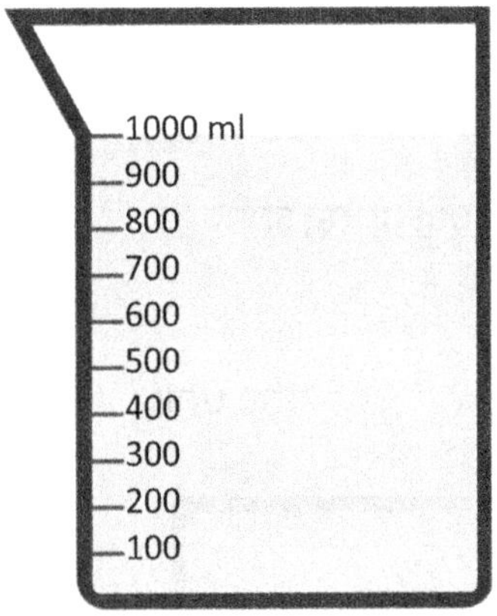

____

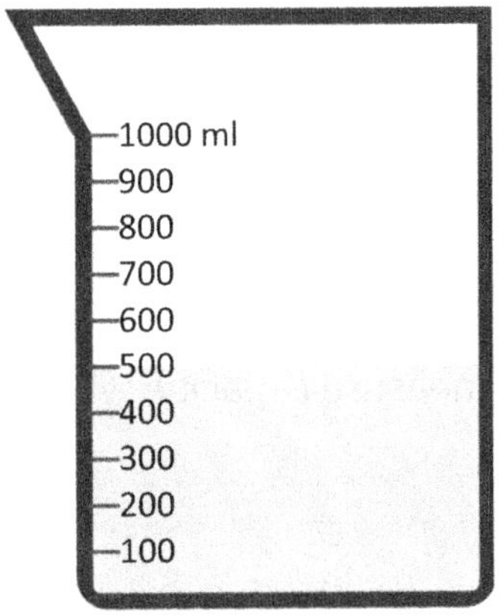

____

5 Leni has been given this medicine by the doctor.

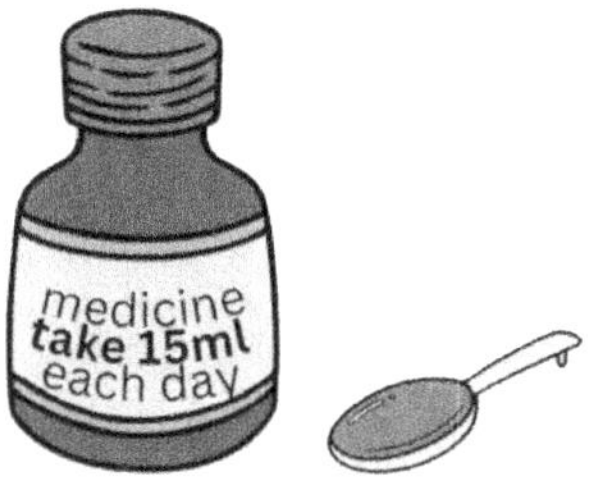

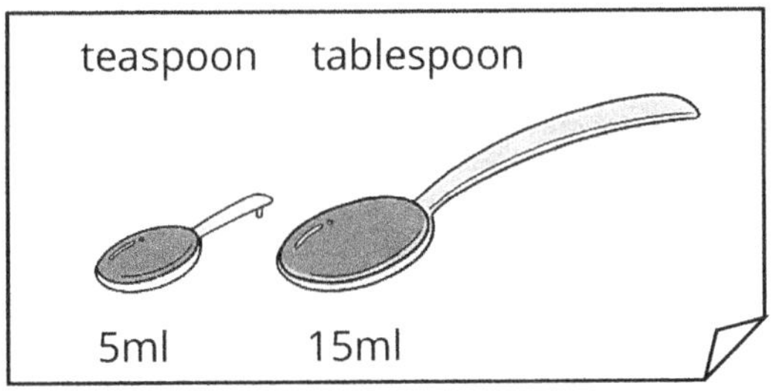

a) How many teaspoons of medicine should Leni take each day? ______________

b) If Leni uses a tablespoon instead of a teaspoon, how many of them should he take? ______________

6 Kelly needs to fill a bucket with water. The bucket can hold up to 12 liters. She has a jug that holds 1.5 liters. How many full jugs of water will she need to fill the bucket completely? You can use a calculator.

____

7 A can of soup contains 0.4 litres of soup. Ann wants to make 1.5 litres of soup.
How many cans of soup does Ann need to buy? Show your working. You can use a calculator.

____

8 a) Jenny has put this much oil in the measuring jug.

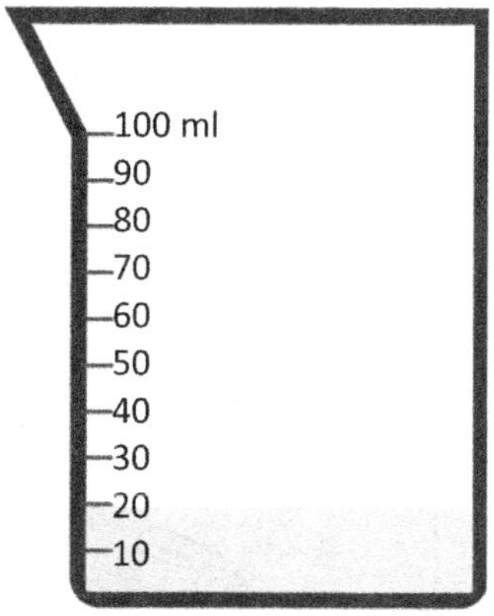

She adds another 70 ml. How much oil is there in total?

____

b) What fraction of the measuring jug has been filled?

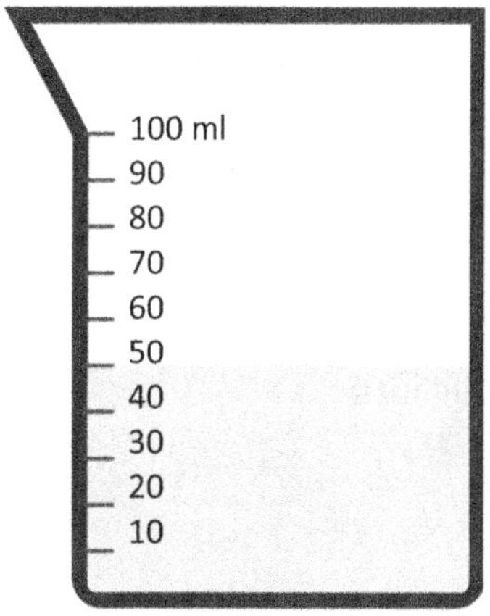

____

c) One bottle of sunflower oil holds 1 litre.

How many litres of oil are there in 12 bottles?

____

d) Hoi plans to drink 2 litres of water each day.

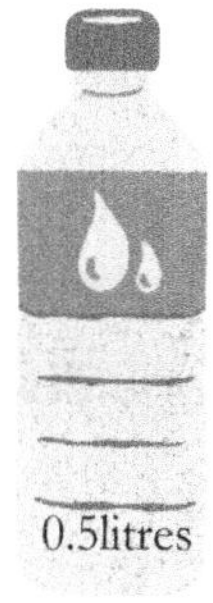

How many of these bottles does she drink each day?

____

e) Bernadette is flying on a plane to go on holiday. She can only take up to 100ml of liquids.

Can they take this sun cream bottle on the plane?

____

f) This container holds 15 litres of water. Mat is filling 2 litre bottles with the water in this container.

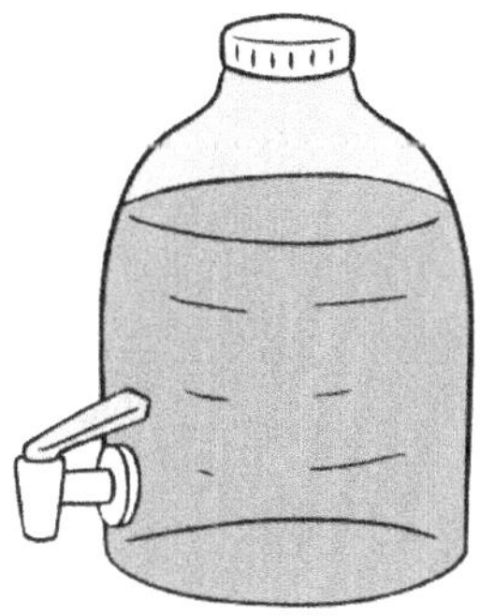

How many full bottles will he be able to fill?

____

9 Charles thinks 12.0 litres is the same as 12 litres. Is he correct?

____

10 Circle the correct option for each item.

a) The capacity of a water bottle is about:

1 millilitre   1 litre

b) The capacity of a bottle is about:

100 millilitres   100 litres

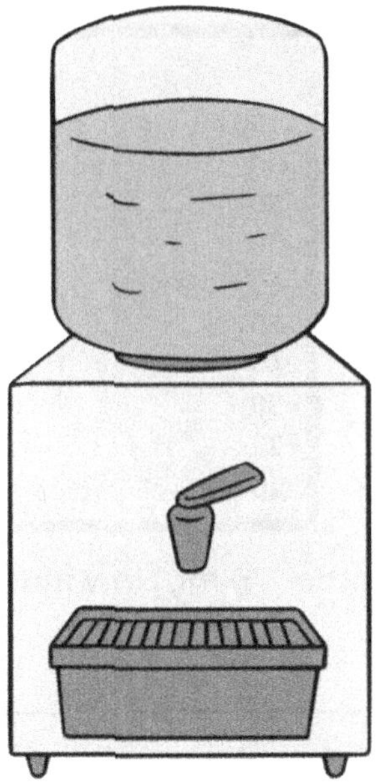

c) The capacity of a water dispenser is about:

10 millilitres   10 litres

To convert ml into l, you **divide the measurement in ml by 1000**, or follow this method:

5000ml into l	Start by writing the conversion fact.	**6500ml into l**	Start by writing the conversion fact.
1l = 1000ml _l = 5000ml x5	Under 1000ml, write 5000ml. 1000 **x 5** = 5000ml, so 1 **x 5** = 5l	1l = 1000ml _l = 6500ml x6.5	Under 1000ml, write 6500ml. 1000 **x 6.5** = 6500ml, so 1 **x 6.5** = 6.5l

To convert l into ml, you **multiply the measurement in l by 1000**, or follow this method:

8l into ml	Start by writing the conversion fact.	**2.3l into ml**	Start by writing the conversion fact.
x8 1l = 1000ml 8l = __ml	Under 1l, write 8l. 1 **x 8** = 8l, so 1000 **x 8** = 8000ml	x2.3 1l = 1000ml 2.3l = __ml	Under 1l, write 2.3l. 1 **x 2.3** = 2.3l, so 1000 **x 2.3** = 2300ml

11 Fill the gaps.

0.1 l = 100 ml	0.2 l = __ ml	0.3 l = __ ml	0.4 l = __ ml	0.5 l = __ ml
0.6 l = __ ml	0.7 l = __ ml	0.8 l = __ ml	0.9 l = __ ml	0.11 l = __ ml
0.25 l = __ ml	0.80 l = __ ml	__ l = 650 ml	__ l = 820 ml	__ l = 641ml
__ l = 947 ml	__ l = 163 ml	__ l = 3940 ml	__ l = 3705 ml	__ l = 4689 ml

12 Gary is making lemonade. He is mixing 70ml lemon juice with 93ml water. How much lemonade is he making? ____________________

13 Put these units in order, starting with the greatest.

7 litres   7 millilitres   70 millilitres   70 litres

__________, __________, __________, __________

14 Kelly is filling a paddling pool with buckets of water. The paddling pool needs 96 litres to be filled fully. Each bucket holds 12 litres. Kelly thinks eight buckets are needed to fill the paddling pool. Is Kelly correct?

____

15 Sara thinks the rainwater tank in her garden holds 100 millilitres. What is wrong with her answer?

____

16 A cylinder has 400 ml of liquid in it. If 350 ml is poured out, how much liquid remains in the cylinder? Show your working.

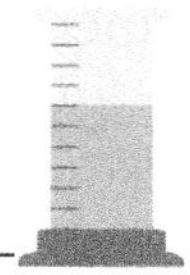

____

17 Tick the container with the smallest capacity.

A - 750 ml   B – 0.7 l   C - 650 ml   D-1.1 l

18 Read the measurements in these scales. Don't forget to add the unit.

a) Write this measurement in two different ways. Hint: using two different units.

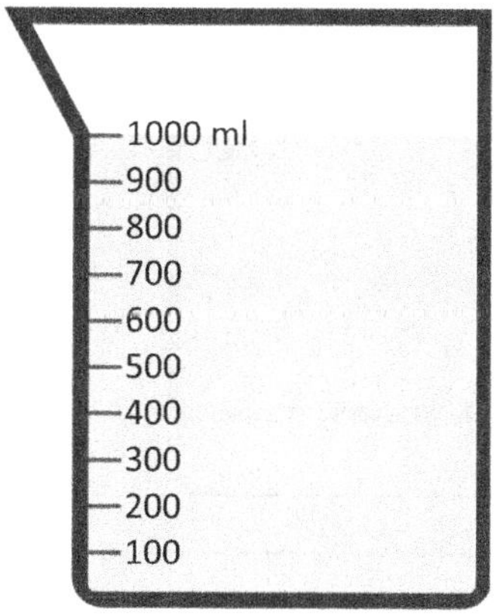

\-   \-

b) What fraction of a litre is this?

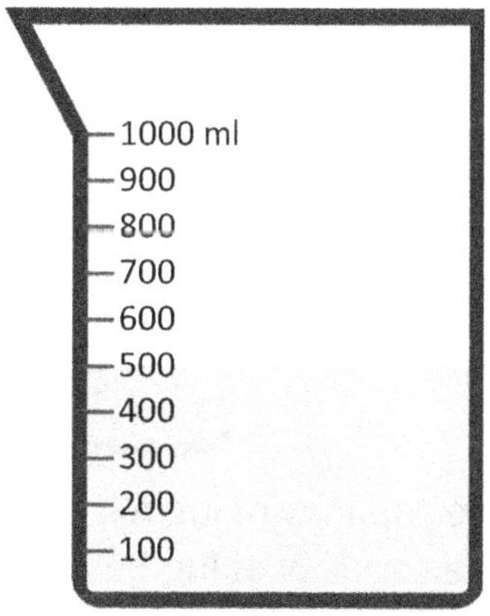

____

19 Zain is going on holiday by plane. He can only take liquids that hold up to 100ml.

A	B	C	D
120 ml	0.1 *l*	0.85 *l*	85 ml

Tick the items that Zain can take on the plane.

20 Which of these is the lowest amount?

1.5 litre	1.05 litre	1.50 litre	1200ml

21 Jeta aims to drink at least 1 litre of water a day. She's shopping for a water bottle that holds at least 1 litre.

A	B	C	D
1000 ml	0.9 l	0.45 l	750 ml

Tick the bottle that Jeta can buy.

22 Jasmine buys coffee. She can get the coffee in different cup sizes. Which cup can hold the most coffee?

A	B	C	D
0.6 litres	400 ml	0.45 litres	550 ml

23 The supermarket sells milk in these different sizes: 2 litres, 1 litre, 750 ml and 500 ml. A customer asks for the carton that holds the most milk. What size carton should the assistant give the customer?

____

Progress check

1 Fill in the gaps below.

7l = _____ ml 5500ml = _____ l

2 Check the capacity of the two jugs below.

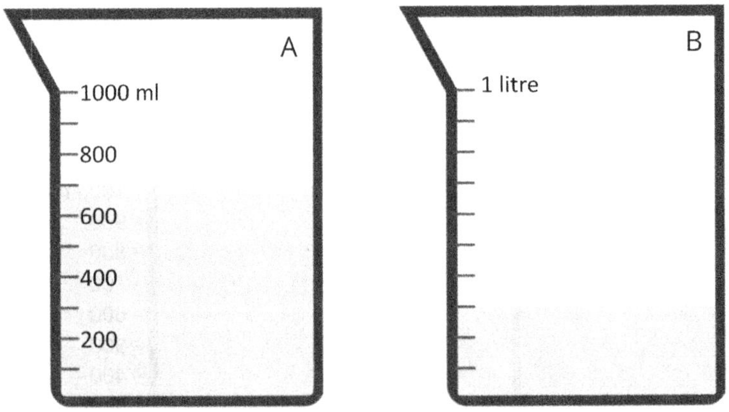

a) What is the capacity of jug A in litres? ________________

b) What is the capacity of jug B in millilitres? ________________

c) What is the difference between the capacity of the two jugs? ________________

3 Put these measurements in order, from the smallest to the largest: 4.5L, 4300ml, 0.45L, 990ml.

__________, __________, __________, __________

4 Sara thinks that there is 610ml of liquid in this container.

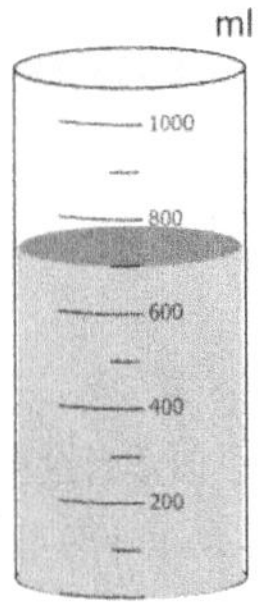

Is she correct? Give a reason for your answer.

____

Use a suitable instrument to measure mass and length

We use these instruments to measure length:

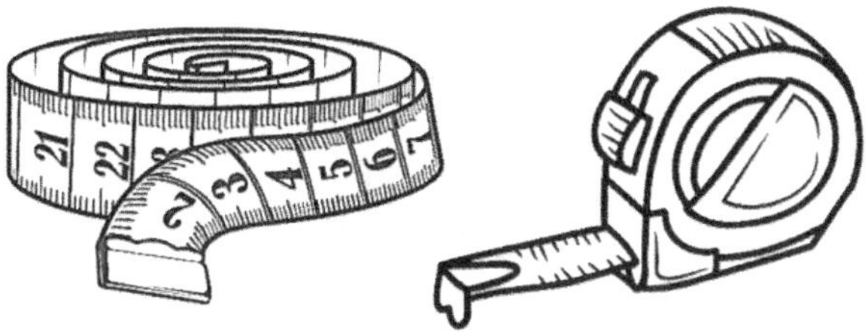

measuring tape or tape measure

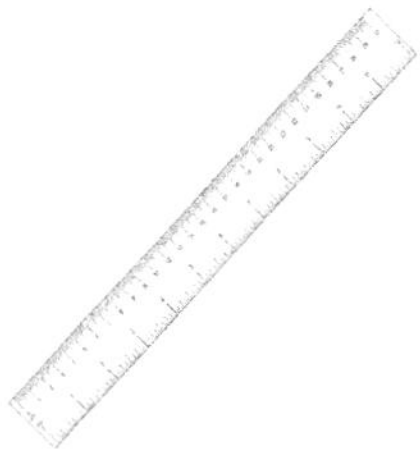

ruler

measuring wheel

We use these instruments to measure mass/weight:

kitchen scale

body scale

luggage scale

Mass means weight

1 Tick the instrument you use to measure mass:

thermometer   measuring jug   kitchen scale   body scale   measuring tape

2 Tick the instrument you use to measure length:

thermometer   measuring jug   kitchen scale   body scale   measuring tape

3 Stef wants to measure the weight of ingredients for making a cake. What measuring instrument should she use?

____

4 Nadia wants to measure the width of her bed. What measuring instrument should she use?

____

5 Anna is planning to make a cake and needs to measure 250 grams of flour. What instrument should she use?

____

6 Thomas needs to measure 2 metres of wire for his school project. What instrument should he use?

____

7 Nick is making a cake. Tick the instrument he will use to measure the ingredients.

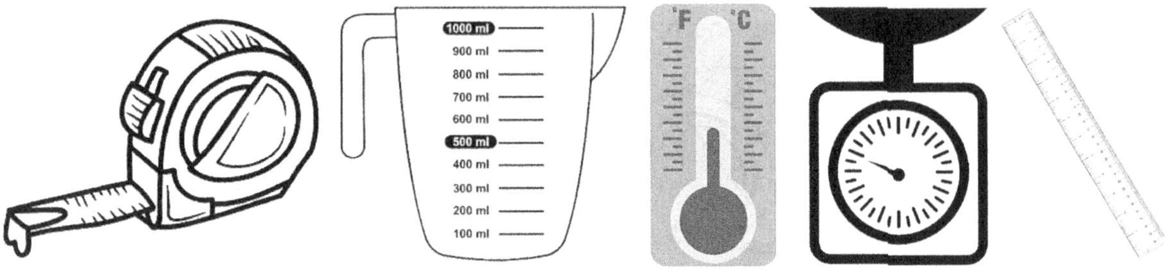

8 Laura needs a new TV unit. Tick the instrument she will use to measure the space where the TV unit goes.

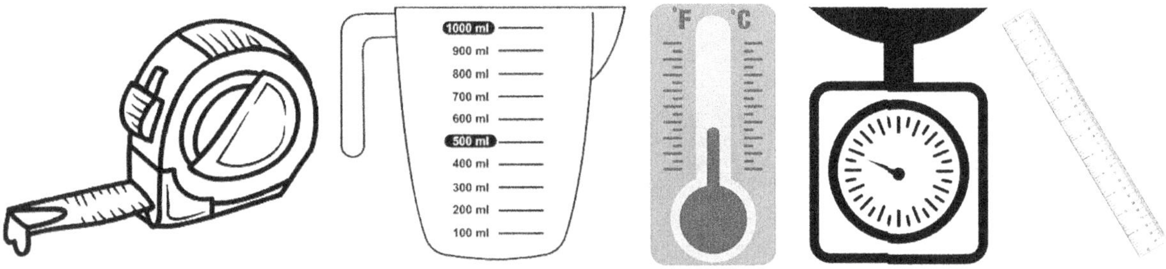

9 Marco is baking bread and needs to measure out 500 ml of water. What instrument should he use?

____

10 Lilly is packing for a trip and needs to measure the weight of her suitcase. Which instrument should she use?

____

Progress Check

1 Name these measurement instruments:

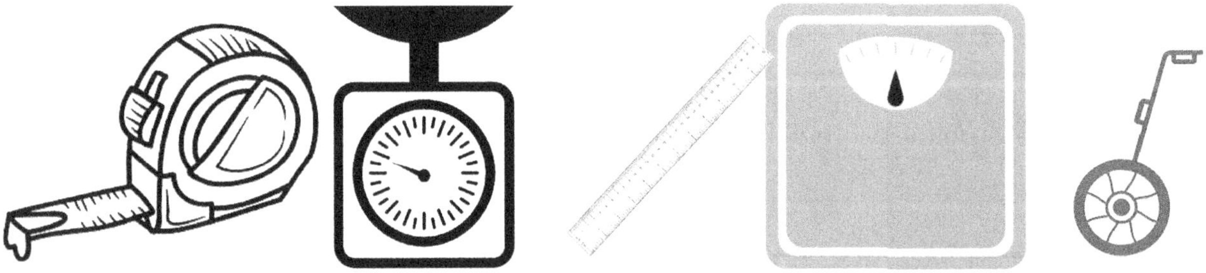

__________ __________ __________ __________ __________

2 What instruments do you use to measure length?

____

3 What instruments do you use to measure mass (weight)?

____

Use and compare measures of length, capacity, weight and temperature using metric or imperial units to the nearest labelled or unlabelled division

We use the divisions and the numbers when measuring:

- length
- weight
- capacity
- temperature

Labelled divisions have numbers in them. Unlabelled divisions don't have numbers.

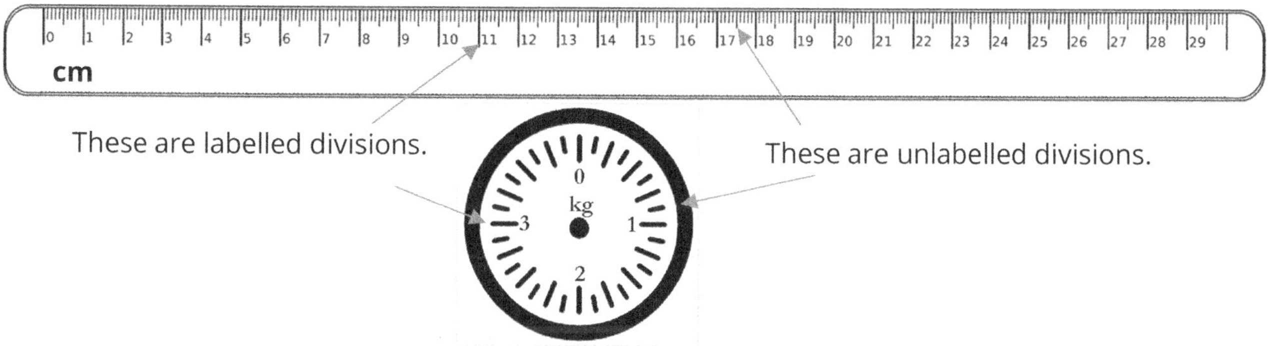

To the nearest labelled division, this line is 8 cm long.

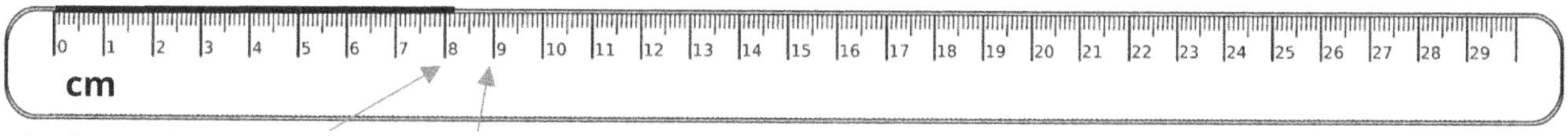

The line is between 8cm and 9cm, but closer to 8cm.

To the nearest unlabelled division, this line is 21.6 cm long.

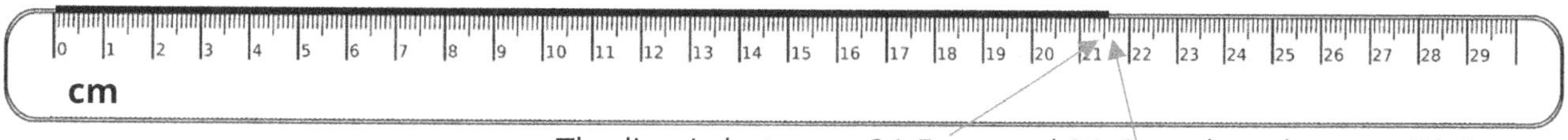

The line is between 21.5 cm and 21.6 cm, but closer to 21.6 cm.

1 Read these temperatures to the nearest unlabelled division.

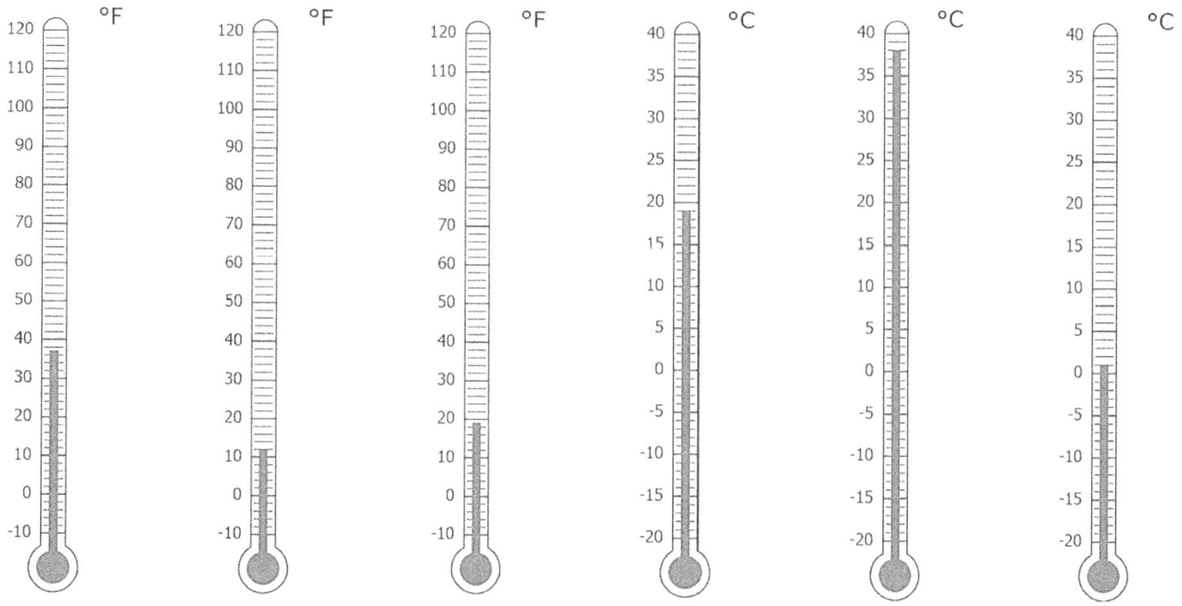

2 a) Tick the thermometer that shows the higher temperature.

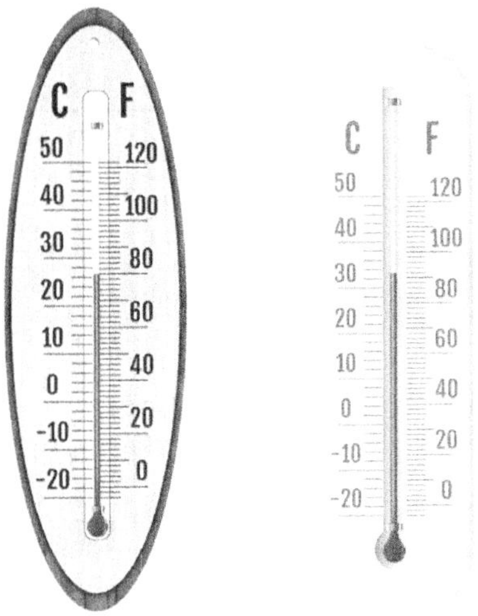

b) What temperature does this thermometer show to the nearest unlabelled division?

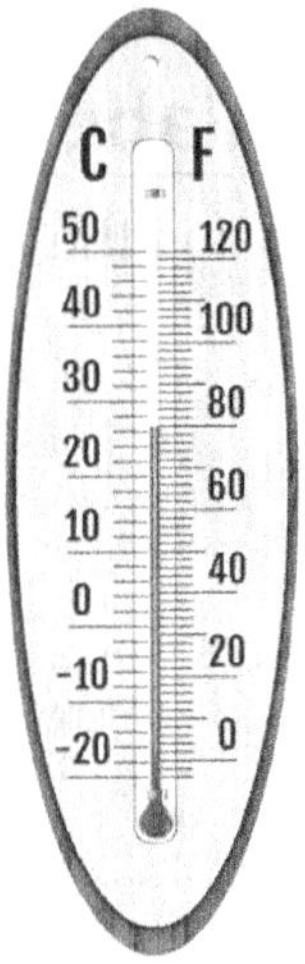

____°C ____°F

3 Nesa thinks this line is 27 cm to the nearest labelled division.

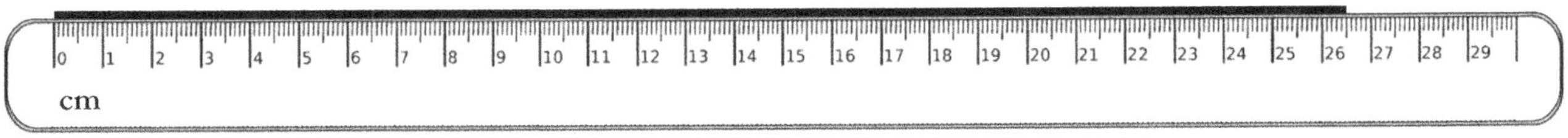

Is Nesa correct? Why?

____

4 How long are these lines to the nearest labelled division?

____cm

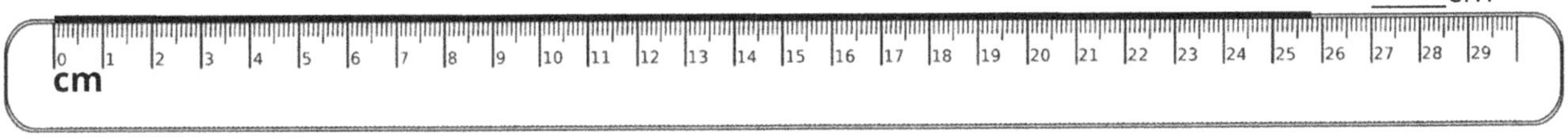

____cm

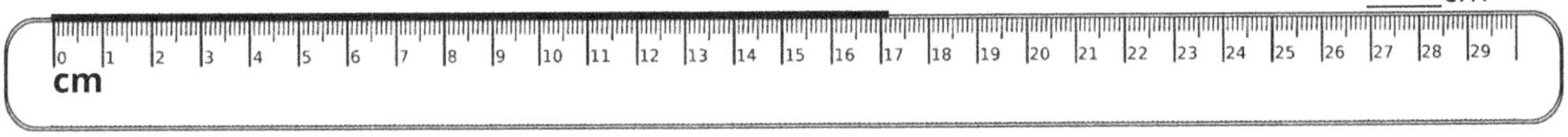

5 Dan thinks this line is 26 mm to the nearest labelled division. Is Dan correct?

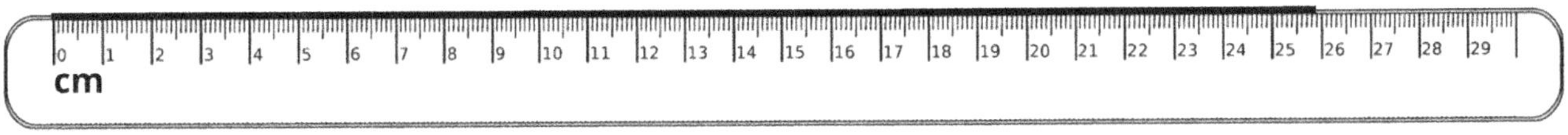

____

6 The length of the wardrobe rounded to the nearest cm is 89cm. What could be the actual length of the wardrobe?

88.5cm, _____, _____, _____, _____, _____, _____, _____, _____, _____

7 How much liquid do these measuring jugs hold to the nearest labelled division?

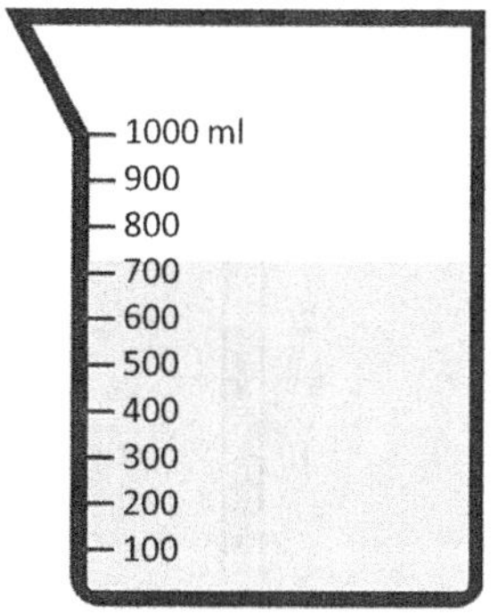

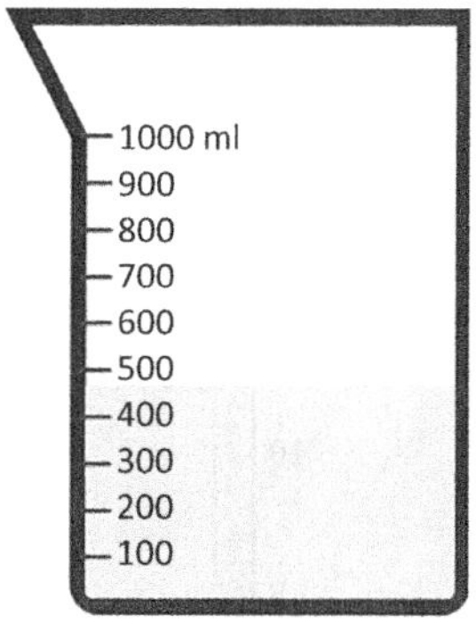

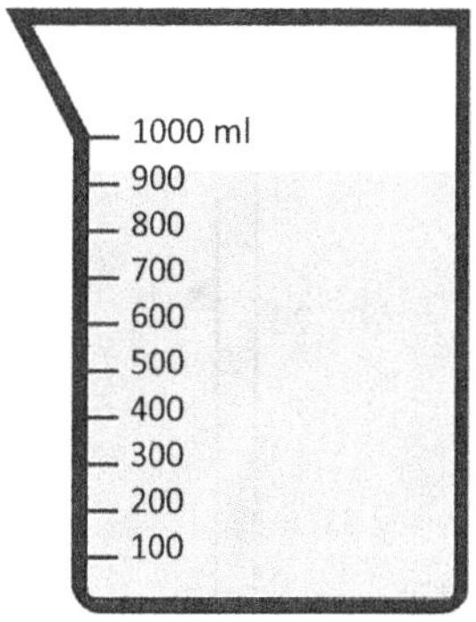

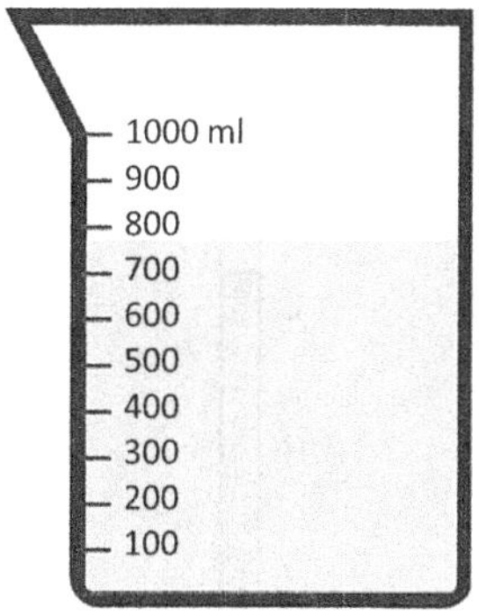

8 What are these weights to the nearest labelled division.

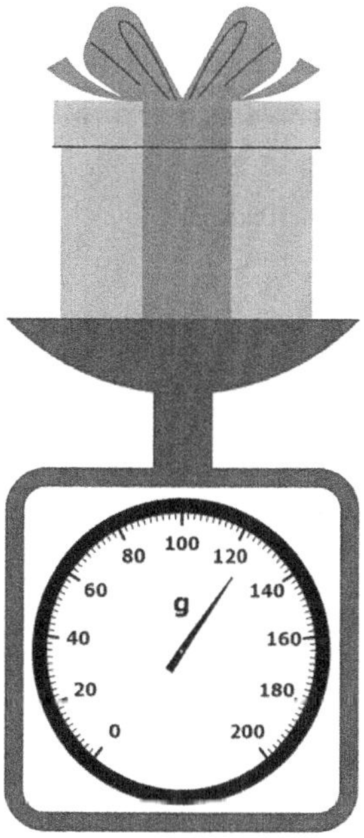

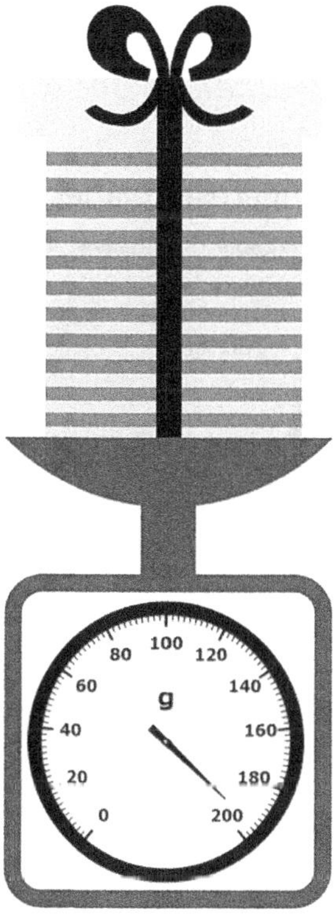

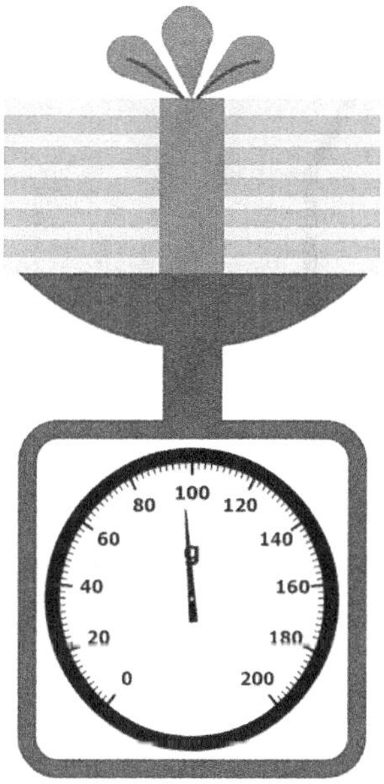

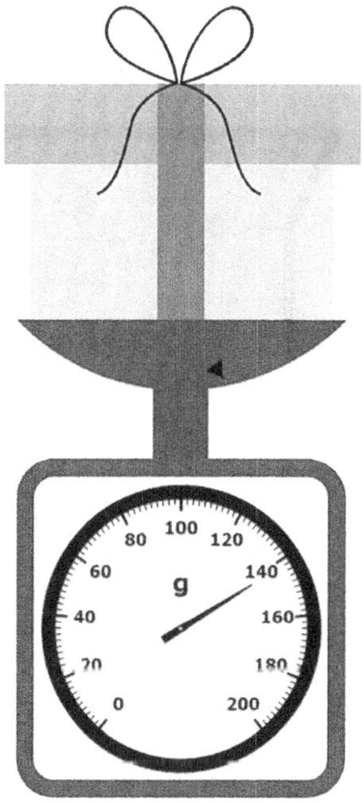

9 a) What weight does this scale show to the nearest labelled division?

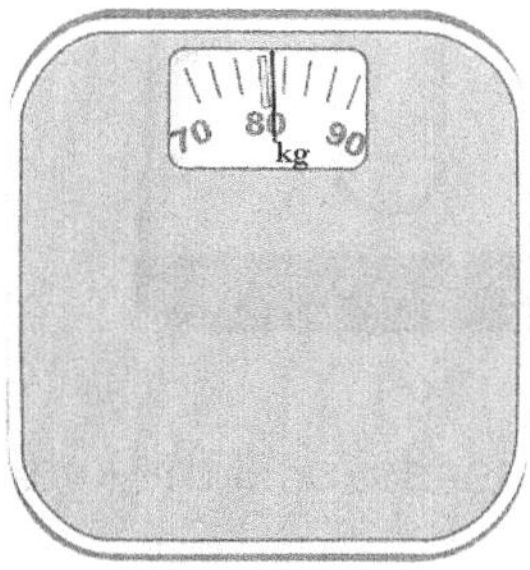

b) Jay thinks this measuring jug holds 1 litre of liquid to the nearest labelled division. Is Jay correct?

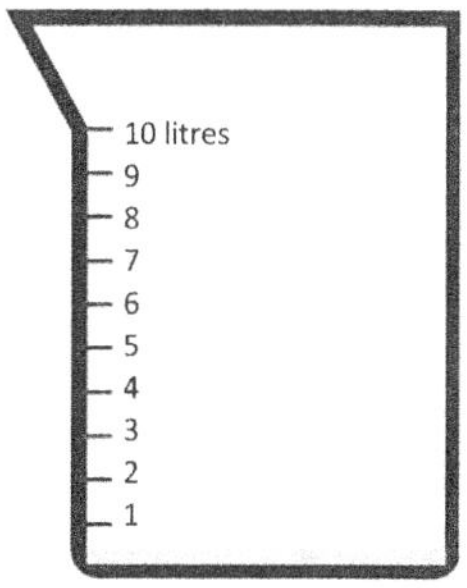

10 Read these temperatures to the nearest unlabelled division.

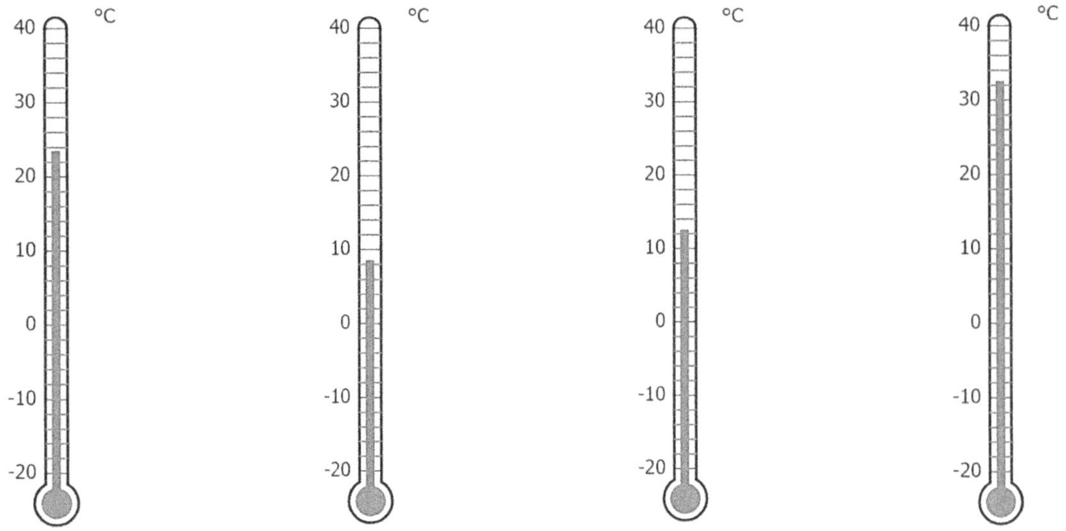

11 How much liquid do these measuring jugs hold to the nearest unlabelled division?

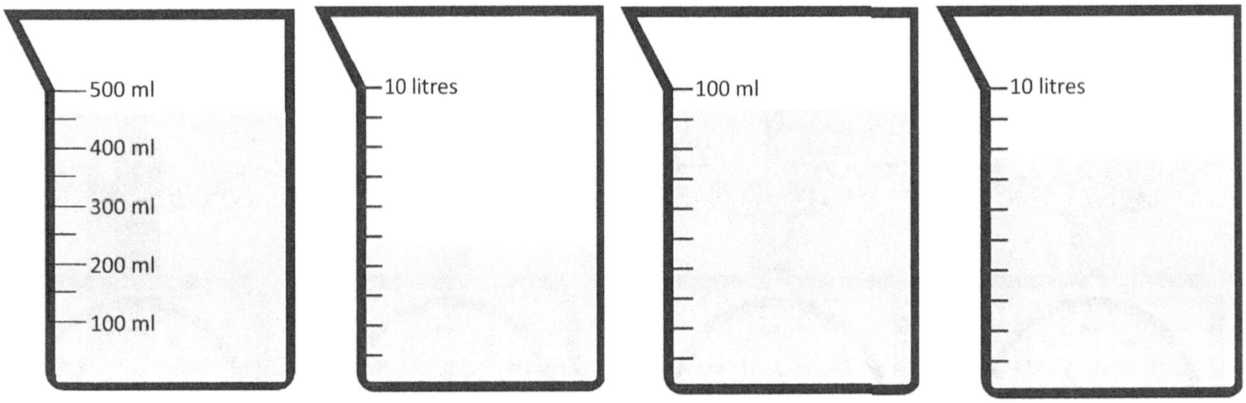

12 a) Ahmed is preparing a cake mix. He pours the milk into a measuring cup as shown below.

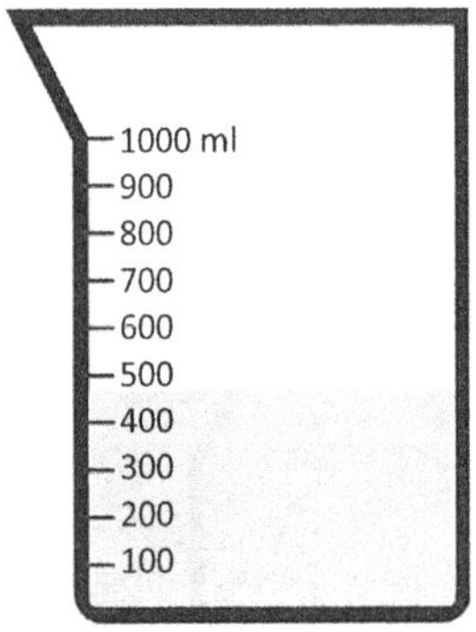

How much milk has Ahmed poured to the nearest labelled division?

b) Luca is cooking. He needs 750 millilitres of boiled water.

Mark the level he needs to fill the kettle up to.

13 Nesa thinks this line is 27 kg to the nearest unlabelled division.

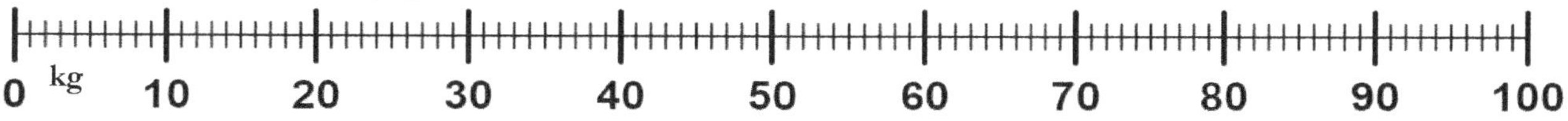

Is Nesa correct? Why? ____________________

Progress check

1 a) How much liquid is there in each of the cylinders? Include the unit.

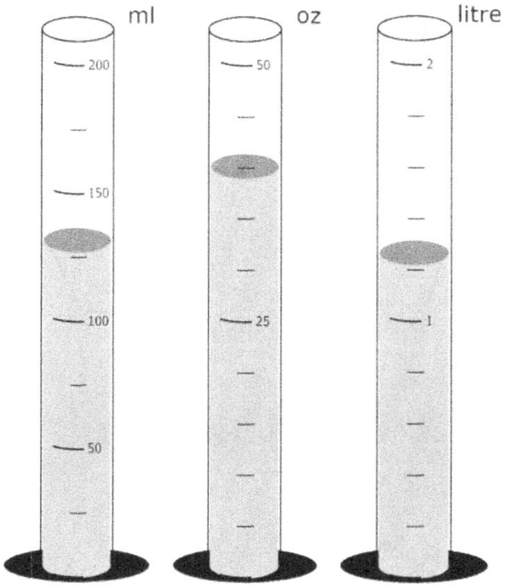

_______ _______ _______

b) Read these measurements to the nearest unlabelled division.

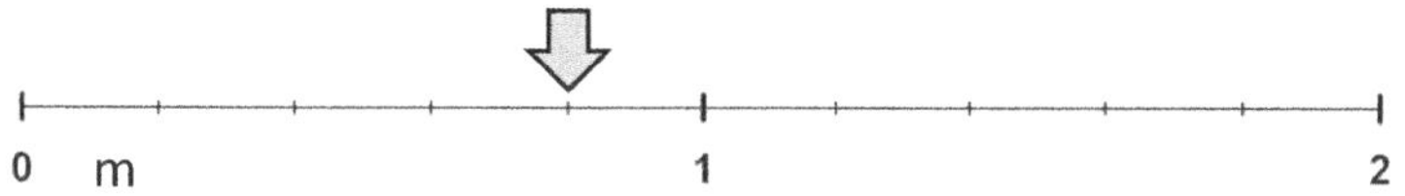

_____ metres

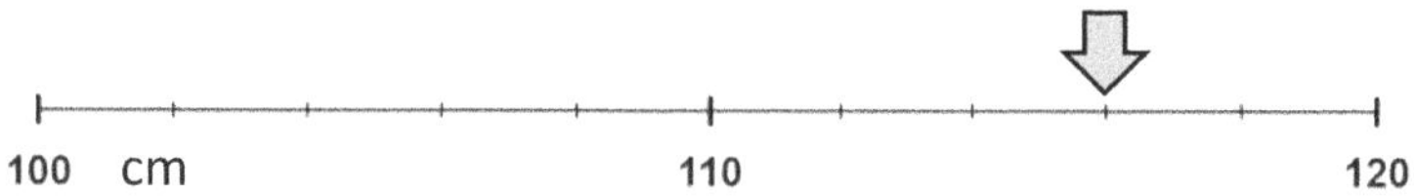

_____ centimetres

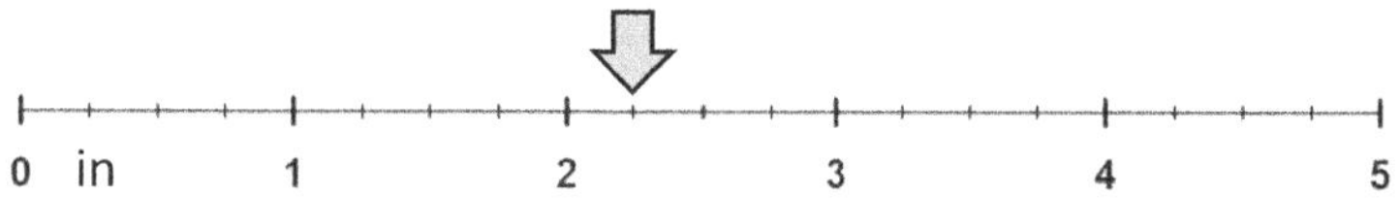

_____ inches

c)i. Write down the temperature each thermometer shows to the nearest unlabelled unit. Include the unit.

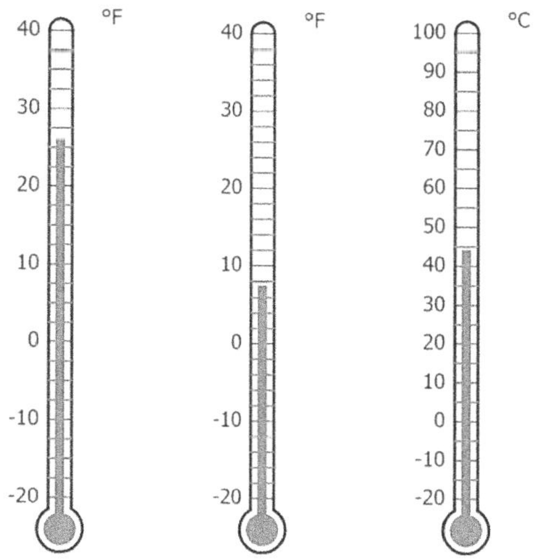

_______ _______ _______

c)ii. What is the difference in temperature between the first two thermometers?

____

d) Read the weight shown in each scale.

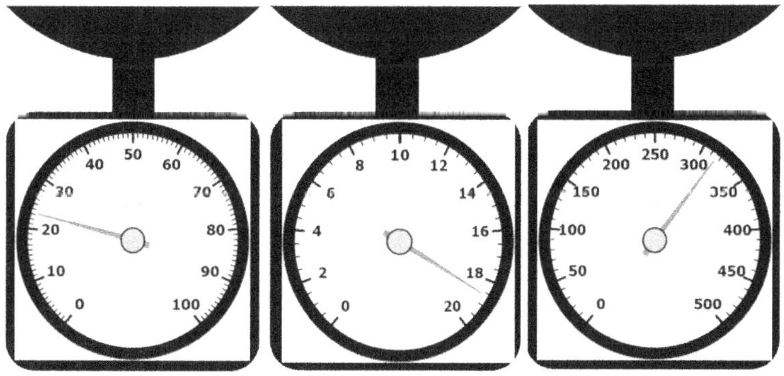

_______ _______ _______

e) What is the length of the pencil to the nearest labelled unit?

___ cm

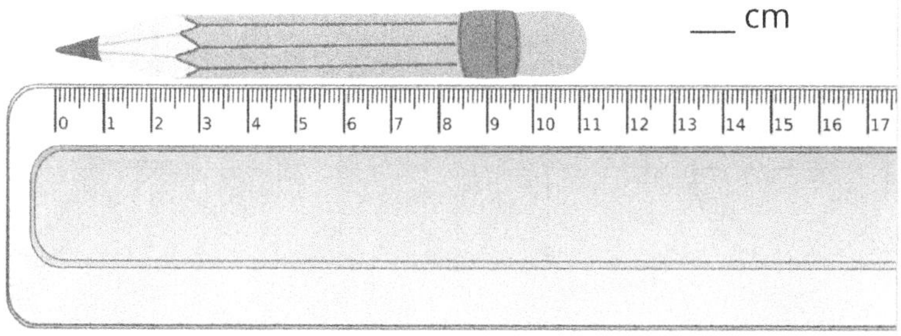

Sort 2D and 3D shapes including using properties including lines of symmetry, length, right angles, angles including in rectangles and triangles

We think of properties of shapes when:

- ordering furniture
- storing food
- doing DIY
- packing for a trip
- gardening
- organizing workspaces

A line of symmetry splits the shape into two identical halves, as if it was a mirror. Here are some examples.

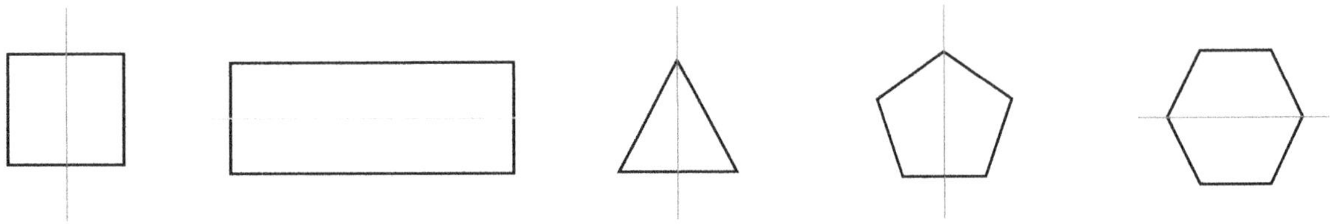

Some shapes have more than one line of symmetry. A rectangle has two lines of symmetry.

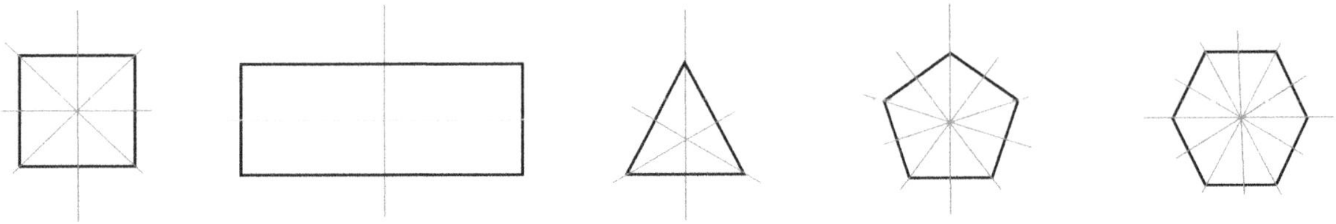

1 Draw all of the lines of symmetry to these regular shapes. Regular shapes have equal sides and angles.

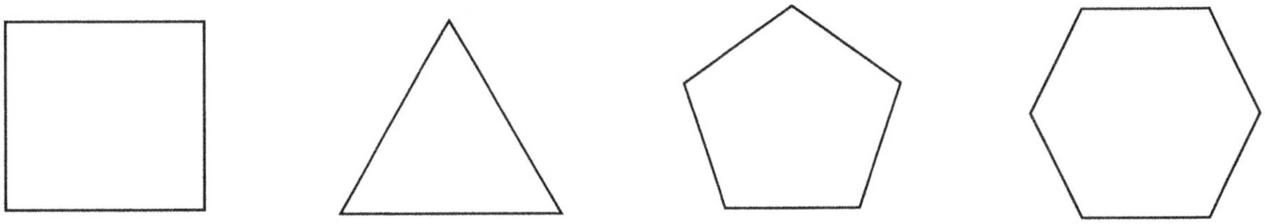

What do you notice about the total number of lines of symmetry in regular shapes?

The number of lines of symmetry in regular shapes is the same as the number of sides/corners.

2 Some shapes have no lines of symmetry. Tick the shape that has no line of symmetry.

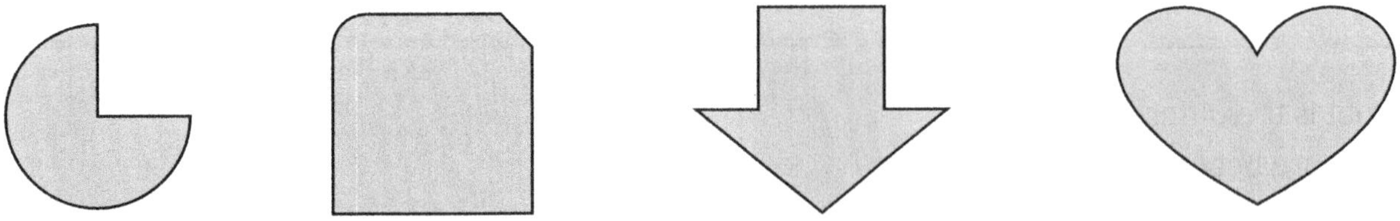

Some shapes have many lines of symmetry. Draw as many lines of symmetry as you can to this circle.

What do you notice?

Circles have an unlimited number of lines of symmetry.

3 a) Write the names of all the 2D shapes you know that have 4 lines of symmetry.

____

b) Write the names of all the 2D shapes you know that have 6 lines of symmetry.

____

A square has 4 right angles. Right angles are the angles that you normally see in the corners of the rooms or corners of a sheet of paper.

This is a right angle.

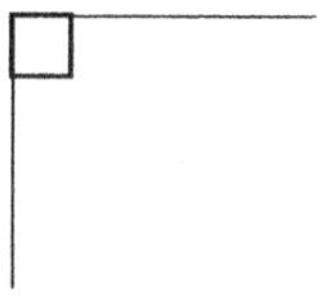

A square has 4 right angles.

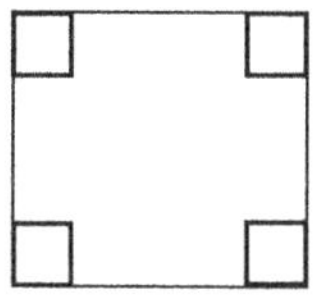

4 Tick the shapes with right angles.

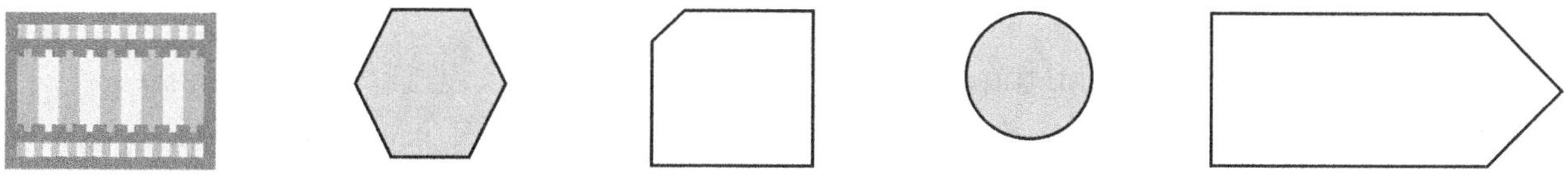

5 How many right angles do these shapes have?

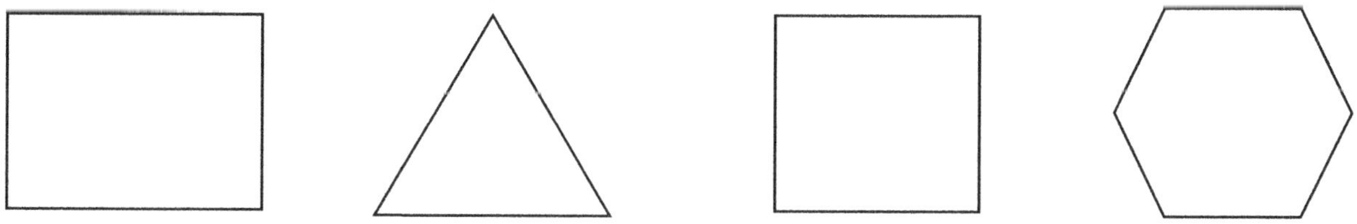

A circle has no corners, so has no right angles.

6 Name the shapes you know that have more than two right angles.

____

7 Tick the shapes that have at least one right angle.

The faces of which of these shapes have right angles? Tick them.

Properties of 2D shapes.

4 sides 4 angles		4 sides 4 angles		3 sides 3 angles	
0 sides 0 angles	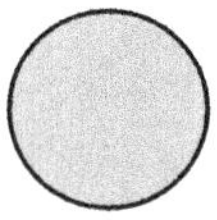	5 sides 5 angles	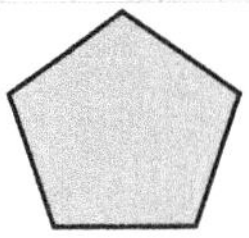	6 sides 6 angles	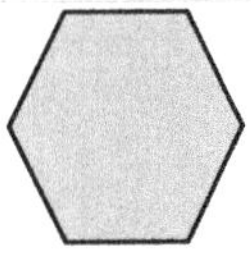

8 a) Write the names of all of the 2D shapes you know that have 4 sides.

____

b) Write the names of all of the 2D shapes you know that have 3 angles.

____

c) Which 2D shape has no sides and no angles?

____

These are the faces, the edges and the base of a 3D shape.

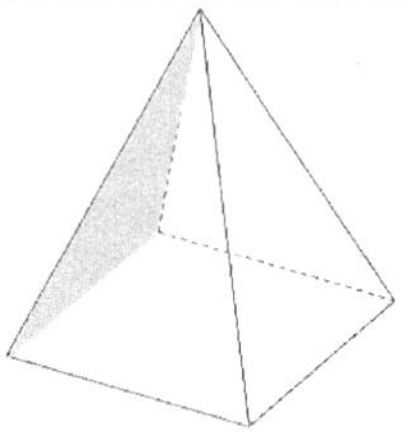	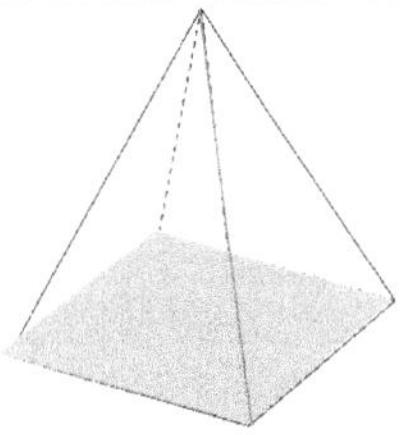	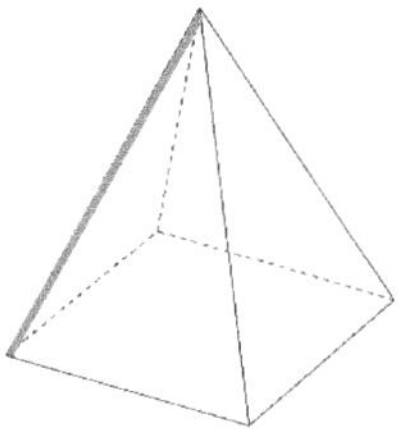
The faces of a pyramid are the 2D shapes you see when rotating the shape. This pyramid has 5 faces	The base is what the shape sits on. This pyramid has a square base.	The edges are the lines where the faces meet. This pyramid has 8 edges.

Properties of 3D shapes.

6 faces 12 edges square base		3 faces 2 edges circle base		6 faces 12 edges rectangle or square base	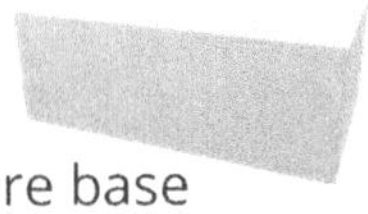
2 faces 1 edge circle base		1 face 0 edges no base		5 faces 8 edges square base	

9 a) Which 3D shapes have 6 faces? ______________________________

b) Which 3D shape has no edges and no base? ______________________________

c) Which 3D shapes have 12 edges? ______________________________

d) Which 3D shape 3 faces and 2 edges? ______________________________

e) How many faces does a cuboid have? ______________________________

f) How many faces does a cylinder have? ______________________________

10 John is building a shed in the garden. The roof of the shed will be in the shape of square-based pyramid. What shape will the floor of the shed be in? ______________________________

11 a) Lorna has bought a new cabinet for the hallway. What shape is the base of the cabinet in?

____

b) Should the cabinet fit into a corner of the hallway?

____

12 How many right angles are there in all of the faces of this shape? ____

Triangles are in four main types.

Right-angled triangle. Has a right angle.	Equilateral triangle. Has equal angles/sides.	Isosceles triangle. Has two equal sides and angles.	Scalene triangle. Has sides of different length.

13 Tick all the triangles with right angles.

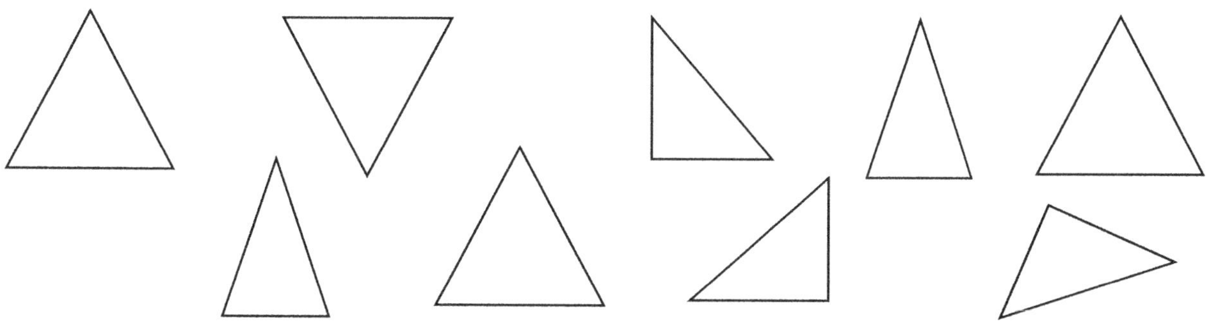

14 Write the type of triangle under each shape.

_______________ _______________ _______________ _______________

15 Draw the lines of symmetry on these triangles.

16 Tick the longest rectangle.

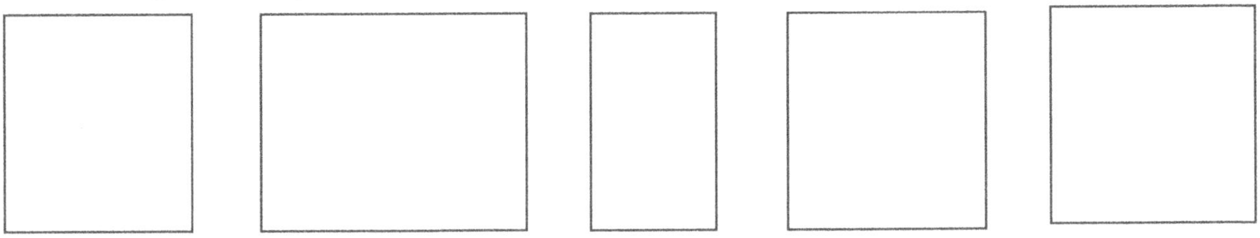

17 Tick the shortest cylinder.

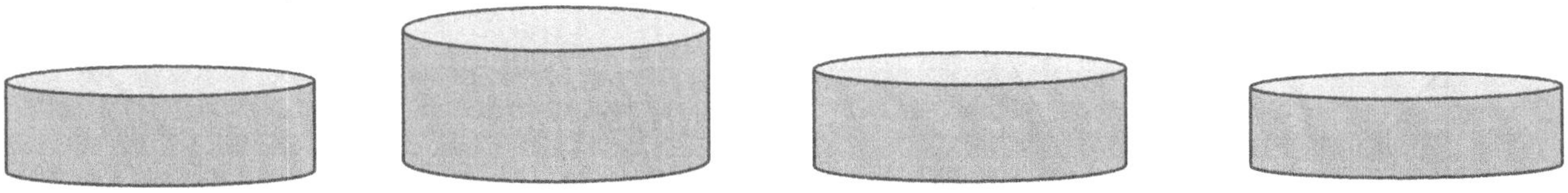

18 Which shapes have right angles at the base? Tick them.

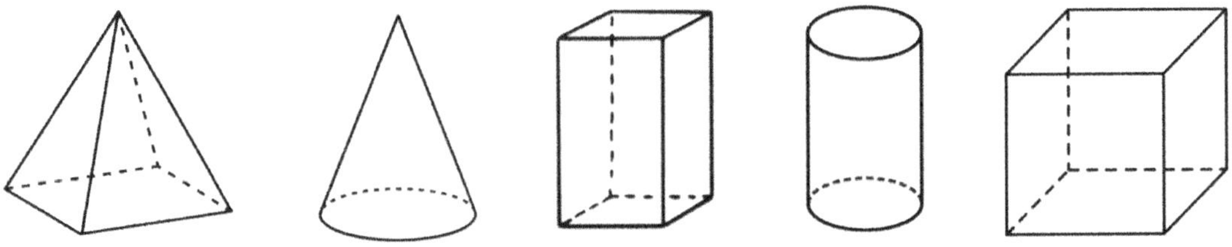

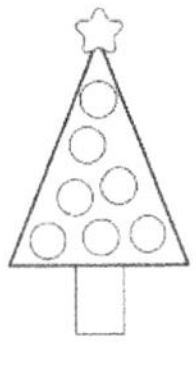

19 a) What type of triangle is this in the tree drawing? _______________________________

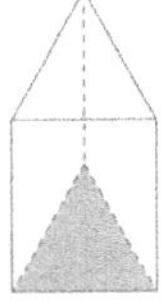

b) The base of this prism is triangular. What type of triangle is this? _______________

Progress check

1 Tick the shapes which have exactly two lines of symmetry.

2 a) What is the name of a shape that has 6 sides? ______________________

b) What is the name of a shape that has 5 sides? ______________________

3 Tick the shapes below that have no angles?

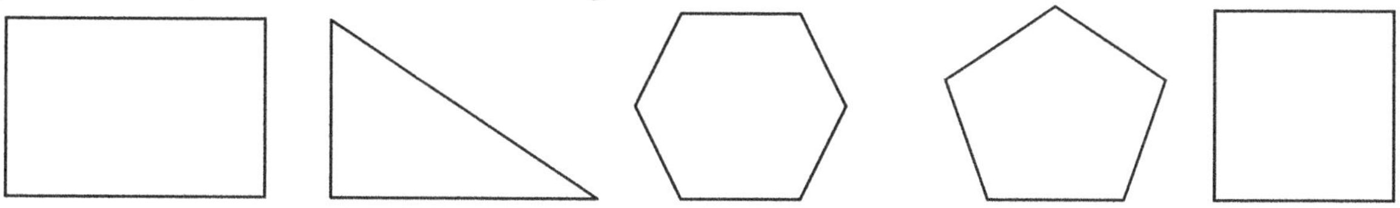

4 a) William thinks that a circle has no lines of symmetry. Is he correct?

____

b) Pretty says, 'Every triangle has three lines of symmetry'. Is this correct? Explain.

____

c) William thinks that a sphere has a circular base. Is he correct?

____

5 Tick the shapes which have exactly 6 faces.

What is the name of the shape that has 5 corners? ______________________

6 Draw:

a shape with a circular base.	a shape with 4 right angles.	a shape with sides of equal length.

Use appropriate positional vocabulary to describe position and direction including eight compass points and including full/half/quarter turns

We use positional vocabulary to:

- describe the position of a building
- give directions to a place or item
- follow instructions
- store items
- enter and exit buildings

These are the main four compass points: north (N), south (S), east (E) and west (W).
This is useful especially when looking at maps.

Through the compass points you can find the cities in the north, south, east or west of a country or areas in a city. Also, you can find places that are in the north, south, east or west of another.
Examples:

Manchester is in the north.
Bournemouth is in the south.
Norwich is in the east.
Bristol is in the west.

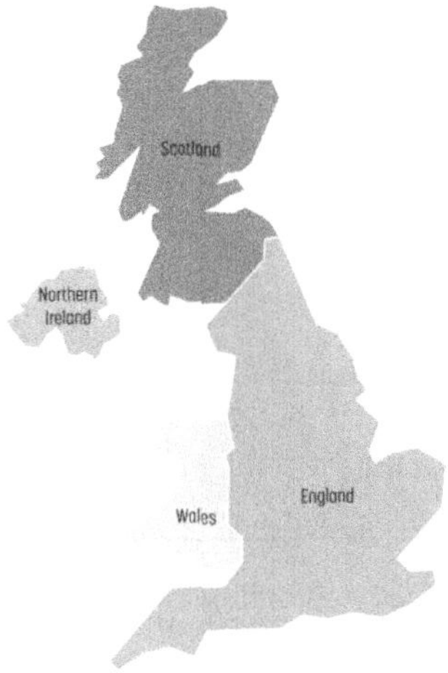

Scotland is north of England.
England is south of Scotland.
Wales is west of England.
Scotland is east of Northern Ireland.

In between the four main compass points, there are another four:
north-west (NW), north-east (NE), south-west (SW) and south-east (SE).

1 Fill these gaps.
North-west is in between ________ and ________ .
North-east is in between ________ and ________ .
_____________ is in between south and west.
_____________ is in between south and east.

Examples:

Oxford is north-west of London.
London is south-east of Oxford.

Exeter is north-east of Plymouth.
Plymouth is south-west of Exeter.

2 Fill the gaps.

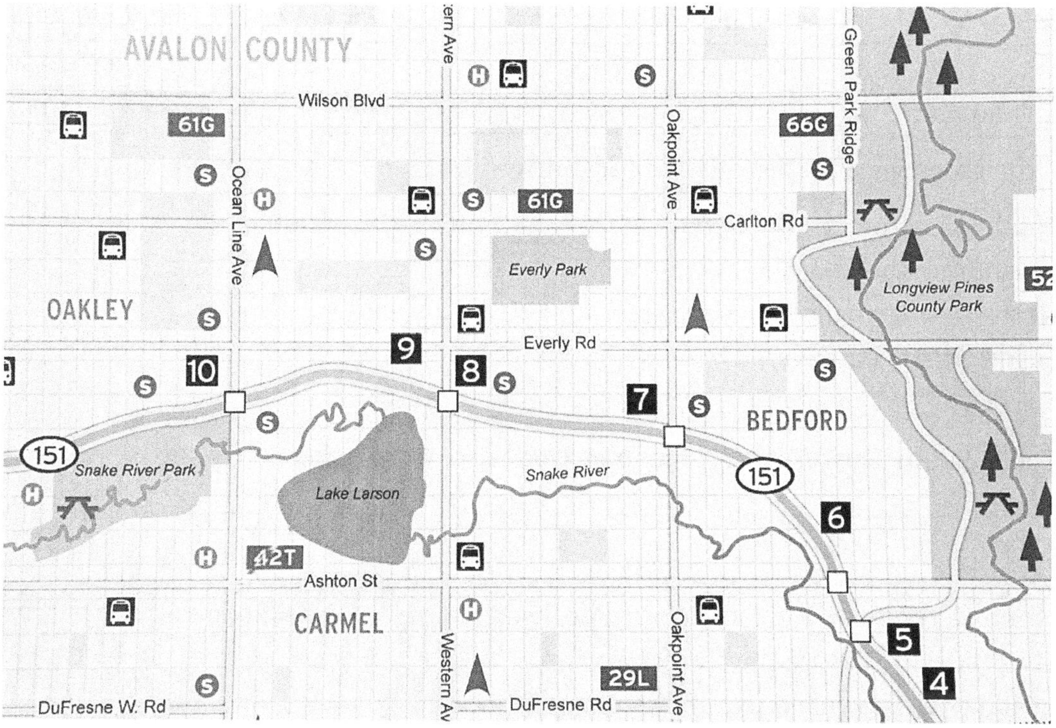

Everly Park is ____________ of Lake Larson. Ashton St is ____________ of Bedford.
Snake River Park is ____________ of Lake Larson. Ocean Line Ave is ____________ of Everly Road.
Green Park Ridge is ____________ of Carlton Road.

To describe position, we also use the terms **turn** and **clockwise/anti-clockwise**.
We should know the terms **full turn**, **half-turn**, **quarter turn** and **three-quarter turns**.

full turn | $\frac{1}{2}$ turn | $\frac{1}{4}$ turn | $\frac{3}{4}$ turn

Clockwise means in the direction the clock moves. Anti-clockwise means in the opposite direction the clock moves.

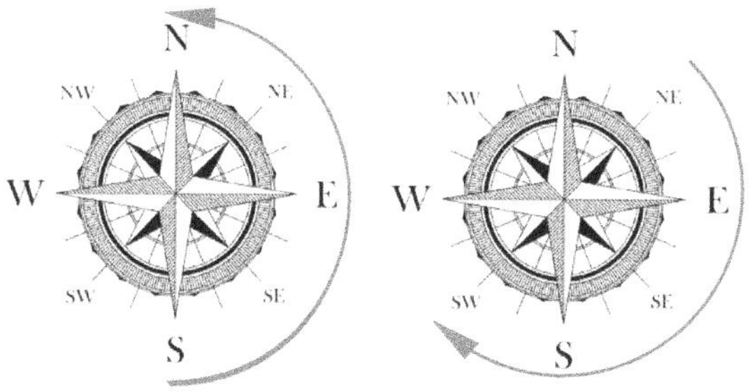

From east to east is a full turn.
From east to west is a half turn.
From east to south is a quarter turn clockwise.
From east to south is a three-quarter turn anti-clockwise.

3 Nina is facing North. She turns a quarter-turn anticlockwise. Which direction is she facing now?

____

4 This is a map of a shopping centre.

Grocery Store	Clothes Shop	Chemist	Card Shop	Café
				Restaurant
		Car park		Supermarket

Where is the grocery store? Circle the correct answer.

A) South-west from the restaurant
B) North from the café
C) North-west from the supermarket
D) East from the Card Shop

5 Dora is facing South. She turns a quarter turn clockwise. Which direction is she facing now?

____

6 Lynn is writing a report in Microsoft Word. She inserts some text upside down.
Lyn needs to turn the picture the correct way up.
What fraction does Lynn need to turn the picture?

____

Progress check

1 Fill in the gaps about the compass on the right.

SW stands for ________________ .	____ stands for **north-west**.
SE stands for ________________ .	____ stands for **north-east**.

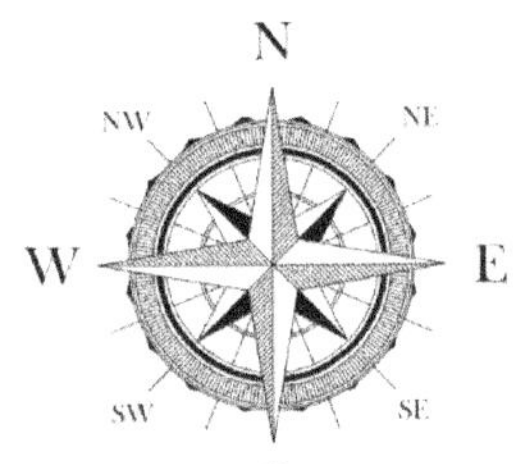

2 Fill in the gaps below, using this UK map for help.

____________ is west of London.
Cambridge is ____________ of Norwich.
Exeter is north-east of ____________ .

3 Complete the sentence below with **full/half/quarter/three-quarterturn** and **clockwise/anti-clockwise** (where needed), so that the sentences are true.

South is ____________ from the north.
South-west is ____________ from the south-east.
North-east is ____________ from the north-east.
South-west is ____________ from north-west.

4 Saima wants to visit the south-west of England. Which places could she go to?

____

5 Aaron thinks that north is a quarter turn clockwise from east. Is he correct? Why?

____

Extract information from lists, tables, diagrams and charts and create frequency tables

We extract information from:

- menus
- receipts
- timetables

Extract information from lists

Greta sells books online. This list shows how many books she sold each year, over three years.

How many books did she sell in year 2020? 129 books.

Books sold	
Year 2020	129
Year 2021	213
Year 2022	366

In which year did she sell 213 books? In year 2021.

1 Nala went shopping at the weekend. This is the receipt she got.

a) How much did the garden tools set cost? ____________

b) What costs more, an outdoor plates set or the garden tools set?

____

c) Which is the most expensive item in the receipt? ____________

d) Which item costs £16.99? ____________

e) How much did all the items cost? Use a calculator to work out the total. ____________

CASH RECEIPT

Description	Price
Garden dining set	£731.00
Outdoor plates set x 2	£41.89
Garden tools set	£23.99
Charcoal barbeque	£55.00
Charcoal 12kg	£16.99
Total	

#854926593559355 9346#

2 a) Nelly has been saving some money during these months.

January: £150
February: £120
March: £135
April: £130
May: £210
June: £130
July: £115

a)i. In which month did Nelly save £130?

____

a)ii. How much did he save in February?

____

a)iii. In which month did Nelly save the most?

____

a)iv. How much did Nelly save over the 7 months?

____

b) A secondary school records the number of students in each year group.

Year 7: 127
Year 8: 131
Year 9: 144
Year 10: 150
Year 11: 165

b)i. Which year group is the biggest?

____

b)ii. How many students are there in year 8 and 9 combined?

____

b)iii. Use a calculator to work out the total number of students in the school.

____

Progress check

1 This is the menu of a takeaway shop called Pizza Place.

Pizza Place Menu

Margherita Pizza	£12.99
Vegetable Pizza	£11.99
Pepperoni Pizza	£14.99
Extra Topping	£0.50
Garlic bread	£1.50

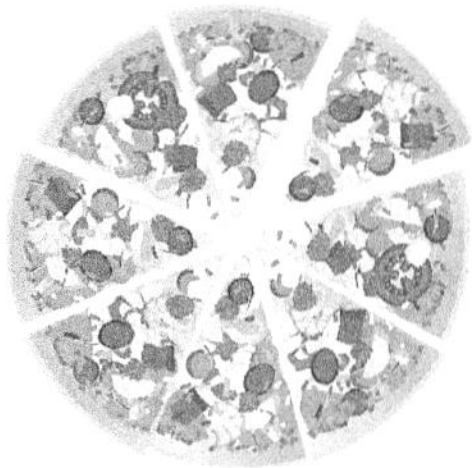

a) How much does a vegetable pizza with three extra toppings cost?

b) Which of the pizzas is the least expensive?

c) How much more than a vegetable pizza does a pepperoni pizza cost?

d)i. Ayla has only £42 in her purse and needs to buy three pizzas and three garlic breads. Which of the pizzas above would she be able to buy?

d)ii. How much change will Ayla be given?

e) Add Milkshake with a price of £2.99 at the bottom of the menu.

2 A company records the number of products sold each week.

Number of products sold

Week 1: 120
Week 2: 150
Week 3: 165
Week 4: 140
Week 5: 180
Week 6: 155
Week 7: 170
Week 8: 160
Week 9: 175
Week 10: 150

a) In which weeks did the company sell more than 160 products?

b) In which week did the company sell exactly 175 products?

c) How many products did the company sell in week 4?

d) In which weeks did the company sell fewer than 150 products?

e) In which week did the company sell exactly 165 products?

f) How many products did the company sell in week 6?

Extract information from tables

Maria wants to buy a shirt from an online store. The available shirt colors and their prices are listed below. Maria wants a blue shirt.

How much does the blue shirt cost?

Colour	Price
Red	£12
Blue	£15
Green	£10

Which shirt costs £10?

1 A college offers these courses.

Subject	Day of the course	Number of enrolments
English	Tuesday	134
Maths	Monday	96
IT	Friday	263
Biology	Wednesday	76
Chemistry	Thursday	42
Physics	Saturday	58

a) What day does the maths course run on?

____

b) What day does the English course run on?

____

c) Which course runs on Thursday?

____

d) Which course runs on Friday?

____

e) How many more enrolments are there for IT than biology? ______________________

f) How many enrolments are there altogether? ______________________

g) How many enrolments are there for the course that runs on Saturday? ______________________

2 The table below shows the distance covered by different coaches from the coach station to different cities.

Destination	Distance (miles)
City A	155
City B	242
City C	168
City D	338

a) Which destination is 168 miles away from the bus station?

____

b) Which destination is 338 miles away from the bus station?

____

c) How many miles further is City D than City A?

____

3 Nora is looking at the weather forecast for next week.

Monday	Tuesday	Wednesday	Thursday	Friday	Saturday	Sunday
18°C	19°C	16°C	22°C	18°C	21°C	20°C

a) What's the temperature on Sunday? ______________________

b) On which day is the temperature 16°C?______________________

c) On which day is the temperature the highest?______________________

c) On which day is the temperature the highest?______________________

c) On which day is the temperature the highest?______________________

Progress check

1 Information about Prevent referral cases has been represented in this table.

Prevent referrals by age in winter 2021-2022	
Age	**Number of case (rounded to the nearest 10)**
Under 15	60
15-20	60
21-30	20
31-40	10
41 and over	10

a) How many 31 to 40 year olds were referred in winter 2021 - 22? ______________________

b) Which age groups had the most of the referrals? ______________________

c) How many more 21-30 than 31-40 year olds were referred? ______________________

d) Work out the actual total number of referrals.

____

2 Sarah has a list of the number of books read each year.
How many books were there in total in these four years?
Show your working.

____

____

Year	Number of books
2020	211
2021	357
2022	452
2023	335

3 Sam tracks his monthly travel expenses. He says his total expenses for January and February were more than £300. Is Sam correct?
Show why you think this.

____

____

____

Month	Expenses in £
Jan	127
Feb	151
Mar	158
Apr	135

Extract information from diagrams

This is an organization chart. Emily is the director of the company. She manages Harper, Brandon and Ava.

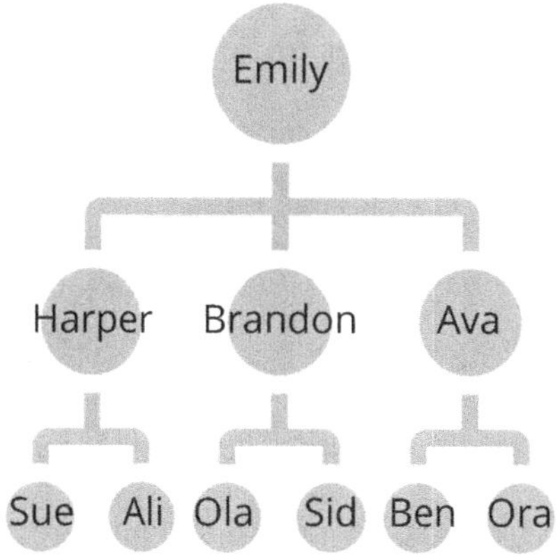

a) Who does Harper manage?

____

b) Who is Ben's manager?

____

1 This is a diagram of a business building.

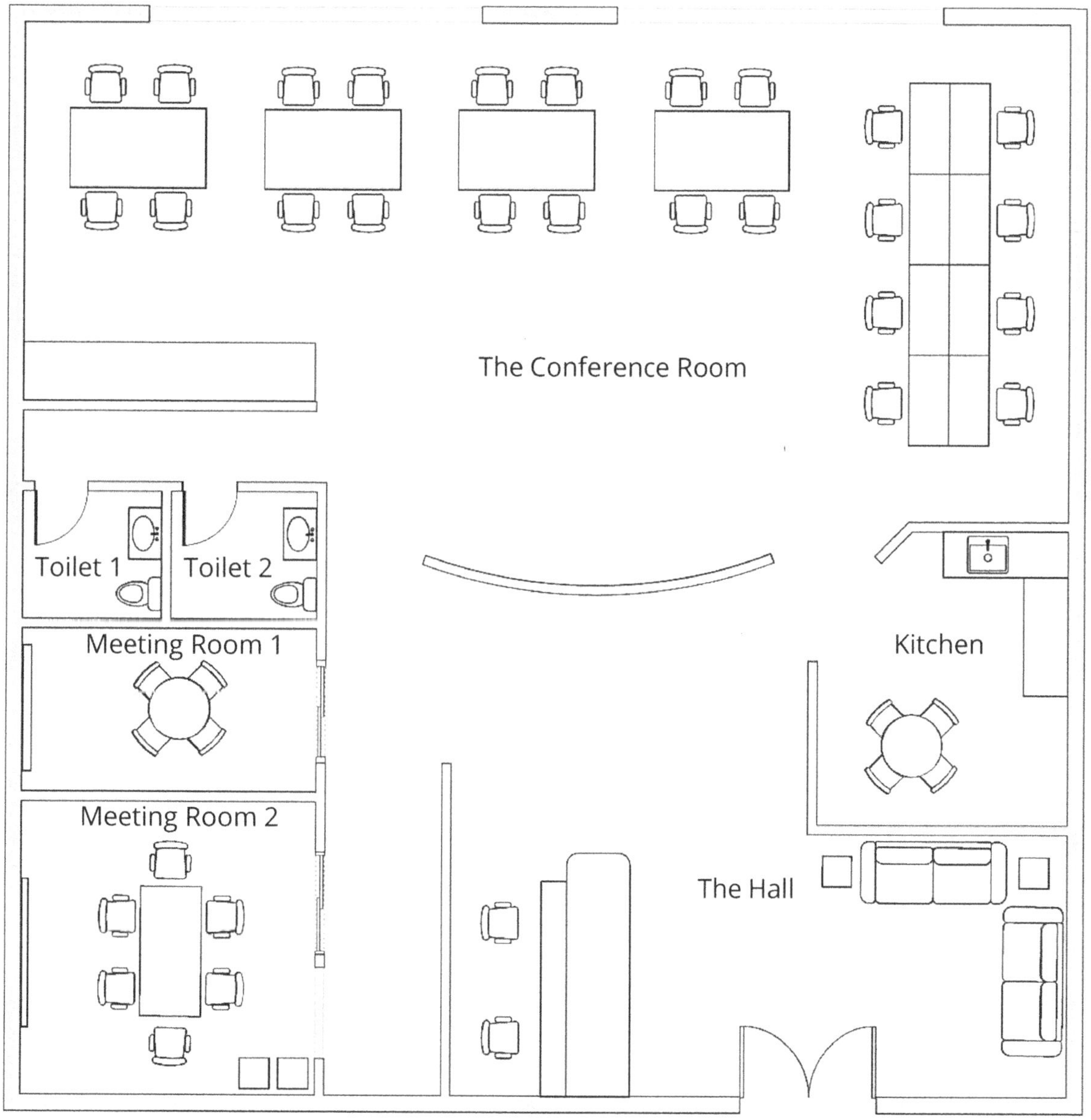

a) How many chairs are there in the conference room? ____________________

b) How many meeting rooms are there? ____________________

c) The kitchen is in between __________ and __________ .

Progress check

1 This is a diagram of some train routes.

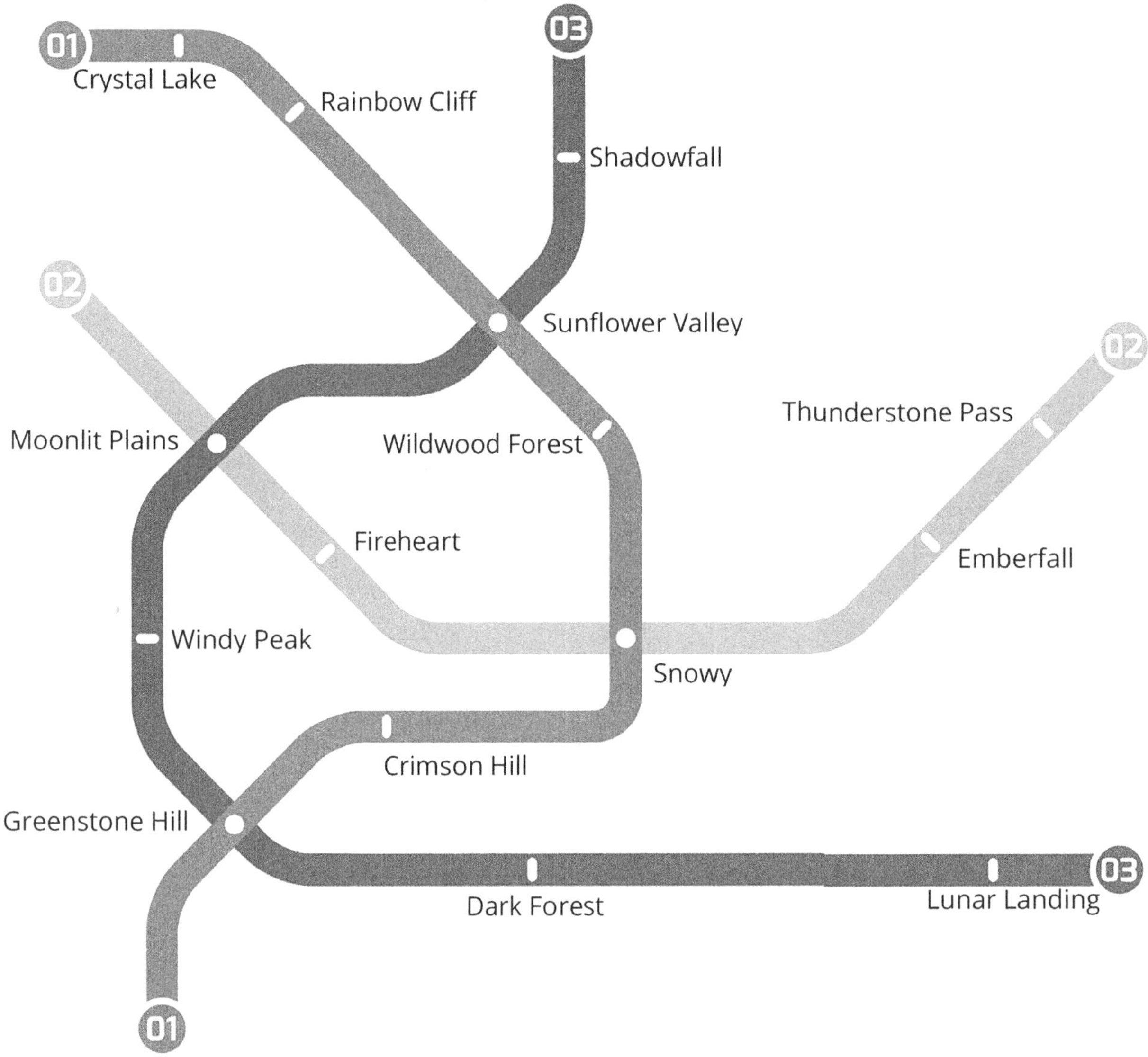

a) Which route is Crimson Hill in? Circle the correct option: Route 01 Route 02 Route 03

b) Isla is travelling from Greenstone Hill to Sunflower Valley. How many stops are there in between?

____

c) Write the names of the stops between Moonlit Plains and Thunderstone Pass.

____

2 This is a food diet that Heather follows.

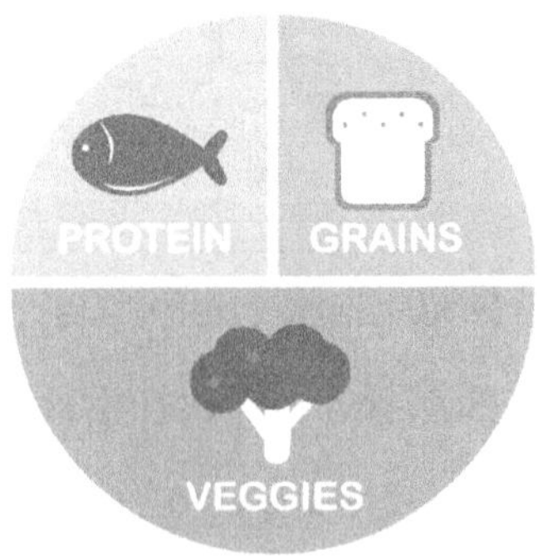

a) What fraction of the whole diet is made of grains?

____

b) Which of the sources of food has the highest portion?

____

Extract information from bar charts

1 This is the number of students in each maths class at MarsCademy College.

This is 26.
Class A has 26 students.

The scale goes up in 1s:
0, 1, 2, 3, 4, 5

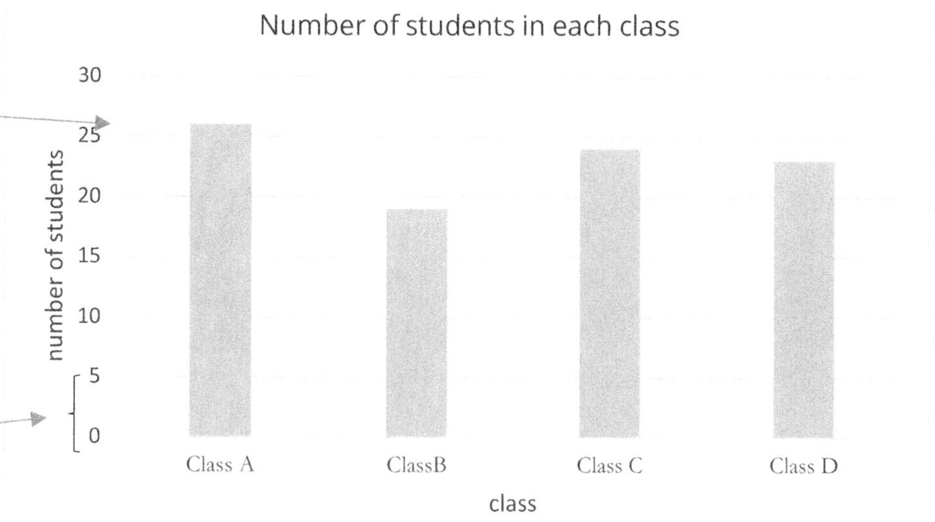

a) How many students are there in Class B? ______________________

b) How many students are there in class D? ______________________

c) Which class has the highest number of students? ______________________

2 This is how much Sunny spent on food last month.

This is £68.
In week 1, Sunny spent £68.

The scale goes up in 2s:
0, 2, 4, 6, 8, 10

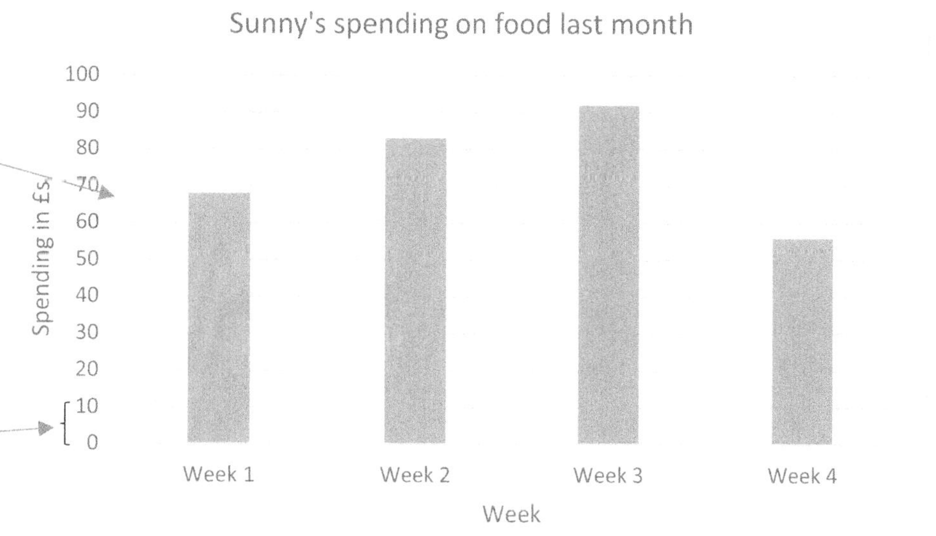

a) How much did Sunny spend in week 3? ______________________

b) How much did Sunny spend in week 4? ______________________

c) In which week did he spend the lowest amount? ______________________

Progress check

1 The bar chart shows the number of cars sold at a dealership each week in February last year.

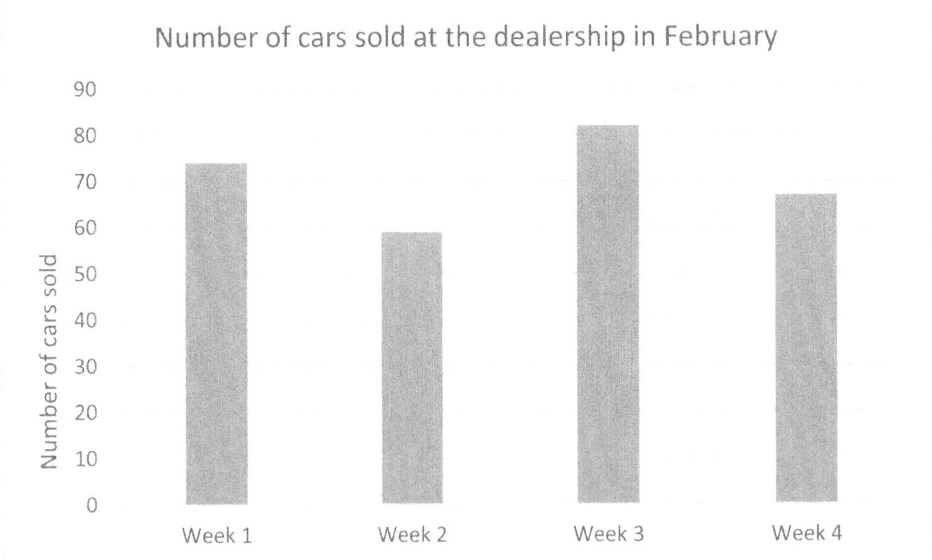

a) Which week had the highest number of sales? ______________________

b) What is the difference in sales between the highest and the lowest week? ______________

c) How many weeks had sales fewer than 70 cars? ______________________

2 The bar chart presents the number of visitors to a museum over seven days.

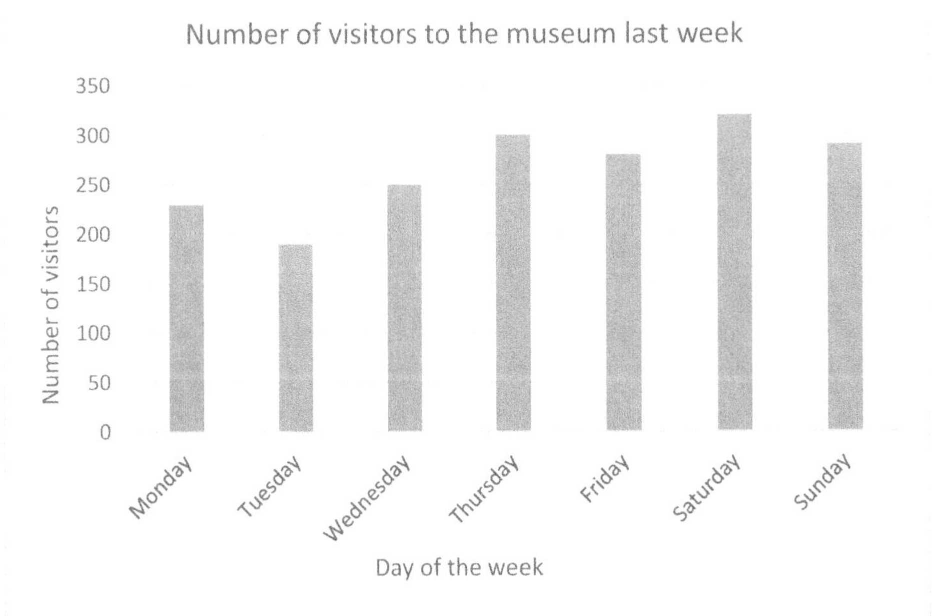

a) On which day did the museum have the highest number of visitors? ______________

b) Calculate the total number of visitors the museum had over the weekend.

____

c) How many more visitors were there on Thursday than on Tuesday?

____

Extract information from tally charts and frequency tables

1 Wadhah asked his students which classes they enjoy more, out of maths, English and IT.

Subject	Tally
IT	||||
English	𝍸
Maths	𝍸 |

| This is called a **tally.**

1 tally means **1 response.**

𝍸 **This** means **5 responses.**

a) How many students chose English? _____ b) How many students chose IT? _____

2 Jackie asked her students how they travel to class. What they said is recorded in this frequency table.

In the frequency column we add the responses in digits.

Subject	Tally	Frequency
walk	||||	4
bus	𝍸 𝍸 |	
train	𝍸 𝍸	
bike	𝍸 |||	

a) How many students take the train to class? ____________

b) How many more students take the bus than cycle? _____

c) Which is the most common mode of transport? _________

d) Fill in the frequency column.

3 Steven works at the dog grooming salon. He has this list of services to be done over a week:
full groom, full groom, full groom, nail clip, bath, bath, nail clip, full groom, full groom, bath, bath, nail clip, bath, bath, full groom, full groom
Complete the frequency table to show this information.

Service	Tally	Frequency
full groom	𝍸 ||	
nail clip		
bath		

4 Sam reads a list of ratings for a movie. Each movie is rated 1, 2, 3, 4, or 5 stars.
These are the ratings: 1, 2, 2, 3, 4, 5, 5, 4, 3, 2, 2, 1, 1, 5, 4, 3, 2, 2, 4, 4 . Complete the frequency table to show this information.

Rating	Tally	Frequency
1		
2		
3		
4		
5		

Progress check

1 A group of learners sold food for charity. Here is what they sold.

Food item	Tally
cupcakes	𝍸 𝍸
milkshakes	||
cookies	𝍸 ||

a) How many cookies were sold? ______________________

b) Which item was the most popular? ______________________

c) How many more cupcakes than cookies were sold? ______________________

2 This is how many fruits and vegetables Anoop had last week.

Day of week	Tally	Frequency
Monday	𝍸 |	
Tuesday	𝍸	
Wednesday		
Thursday	𝍸 ||	
Friday	𝍸 |	
Saturday	𝍸 |||	
Sunday		

a) Anoop had 7 portions of fruit of vegetables on Wednesday. Add this information to the frequency table.

b) Anoop had 6 portions of fruit and vegetables on Sunday. Add this information in the tally chart.

c) Fill in the frequency column of the table.

3 A teacher keeps a record of the different types of books borrowed by her students over the school year. Complete the frequency table for these books: novels (25), non-fiction (15), comic books (20) and biography (10).

Type of book	Frequency

Interpret information, to make comparisons and record changes, form different formats including bar charts and simple line graphs

1 Colleagues at work are voting on whether they should work from home. These are the results.

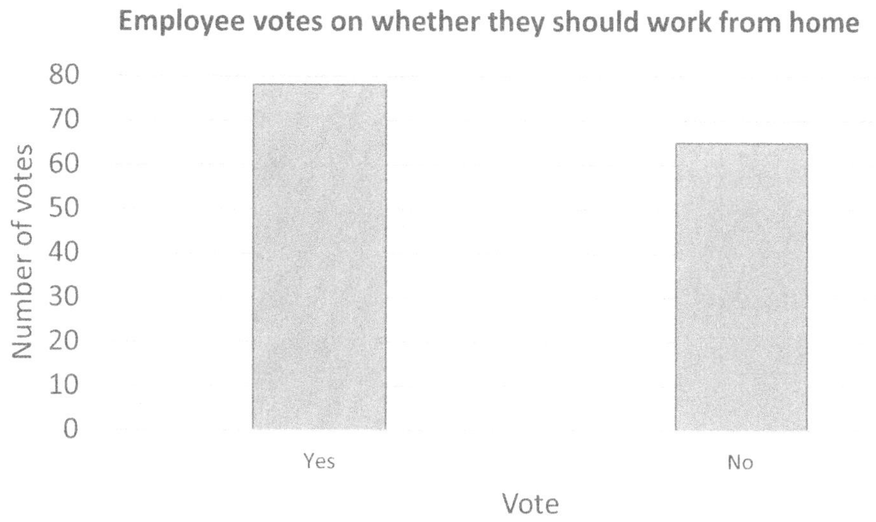

a) How many colleagues prefer to work from home? ______________________

b) Which is the least favourite option? ______________________

c) How many more prefer to work from home than not? ______________________

2 Shobna has her own business. This is how many items she sold from January to April last year.

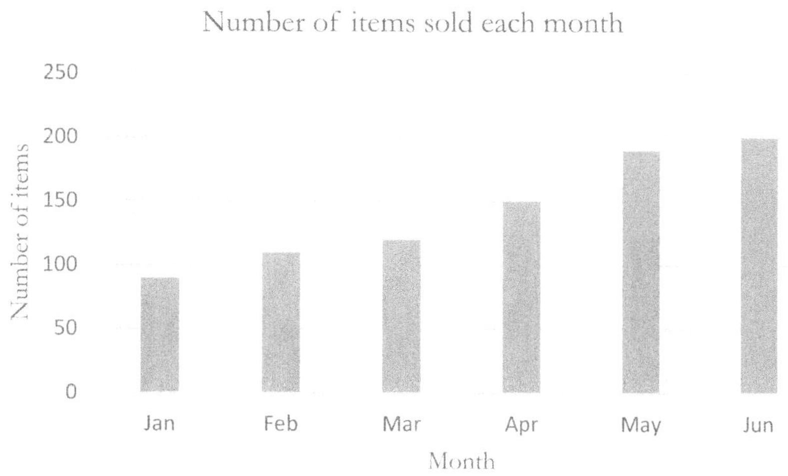

a) How many more items did she sell in March than in February? ______________________

b) In which months did she sell less than 150 items? ______________________

c) In which month did she sell the least number of items? ______________________

d) What trend in sales do you notice? ______________________

3 This bar chart shows how many books Gary read each year from 2020 to 2022.

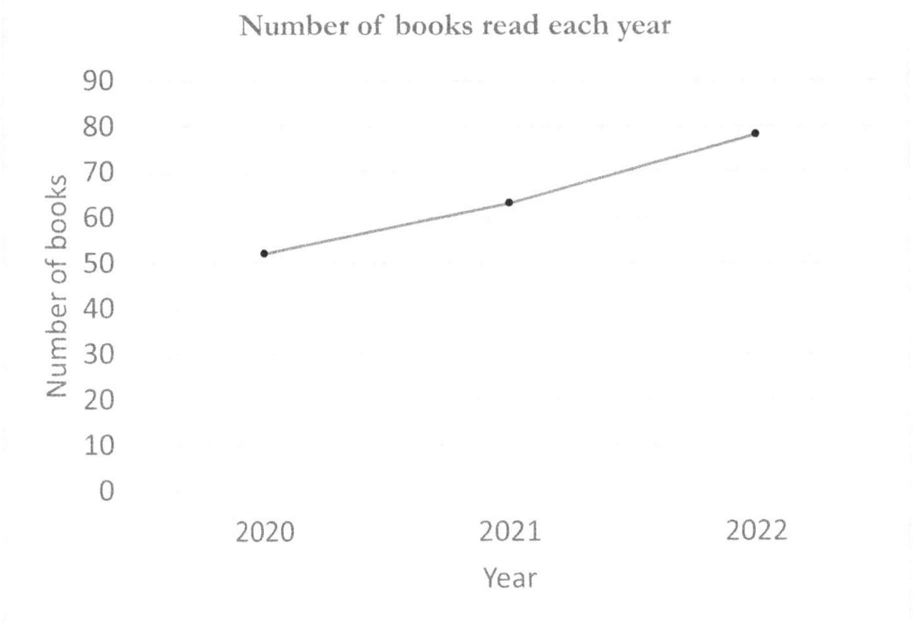

a) In which year did Gary read the lowest number of books? ______________________

b) How many more books did Gary read in 2021 than in 2020? ______________________

c) In which years did Gary read more than 60 books? ______________________

d) Comment on the trend in the number of books Gary has read from 2020 to 2022?

____

4 This is how much rubbish was recycled in a recycling centre last year.

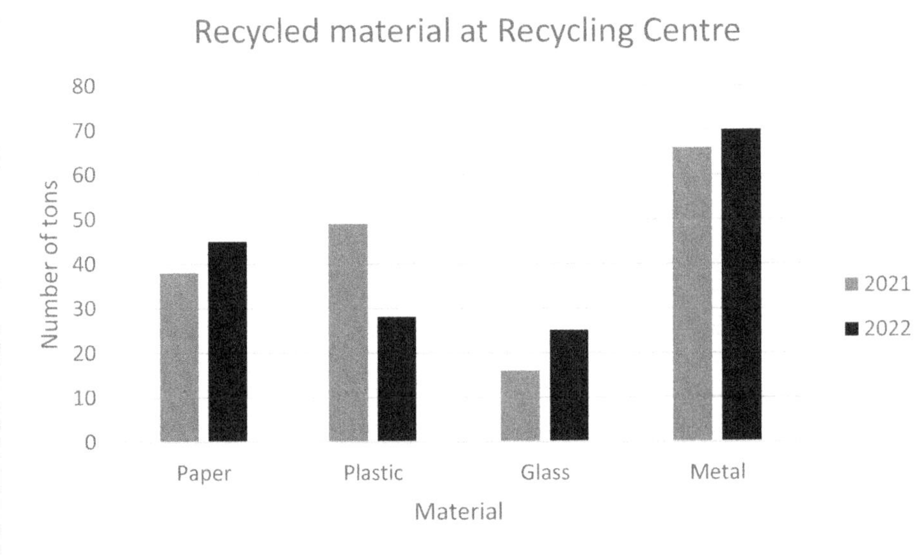

a) Which material is the most recycled one? ______________________

b) How many more tons of plastic were recycled in 2021 than 2022? ______________________

c) Which material has less than 20 tons of recycling material? ______________________

Progress check

1 The bar chart shows the number of books sold each year at Novelty Reads Bookstore from 2018 to 2022.

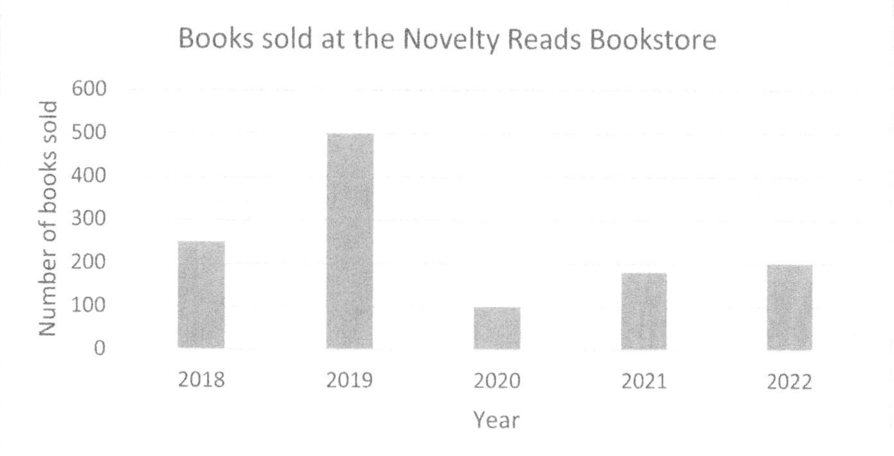

a) In which year did Novelty Reads see the highest increase in book sales from the previous year?

b) Compare the sales trend from 2018 to 2020. What can you infer about the sales performance during this period?

c) How many more books were sold in 2022 than 2021?

2 The line graph shows the average weekly temperatures in Celsius for the month of July in Cedar Town.

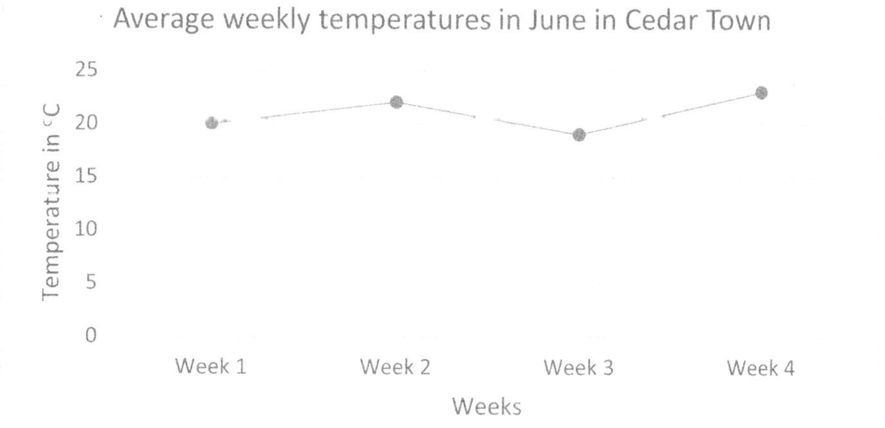

a) During which week did Cedar Town experience the greatest change in average temperature compared to the previous week?

b) How did the temperature trend over the course of the month?

c) By how many degrees Celsius did the average temperature in week 4 increase from week 3?

Organise and represent information in appropriate ways including tables, diagrams, simple line graphs and bar charts

We use charts and diagrams to help us make decisions.

Wadhah asked the students in Class K about the subjects they want to study next year, out of construction, hairdressing and business. He collected the responses in a tally chart.
Wadhah shows the information collected to his colleagues in a bar chart.

The title explains what the bar chart is about.

Subject	Tally
construction	𝍸 𝍸 |
hairdressing	𝍸 𝍸
business	𝍸 𝍸 |||

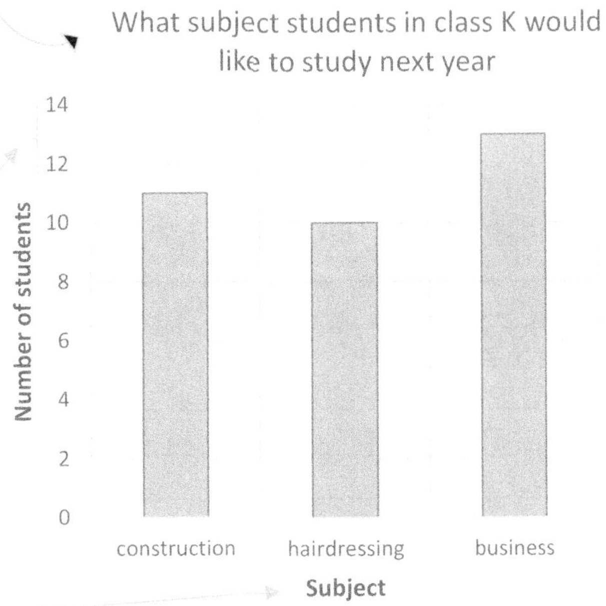

The scale goes up in twos, to 14. Business has the highest (13 students) and 14 covers that.

The axes have been labelled to tell the reader what the numbers and construction, hairdressing and business mean.

1 Jackie is a teacher at MarsCademy College. She is teaching a lesson on the environment and the carbon footprint. She collected information about how students in MarsCademy College travel to college.

Mode of travel	Number of students
walk	20
bus	13
train	15
bike	18

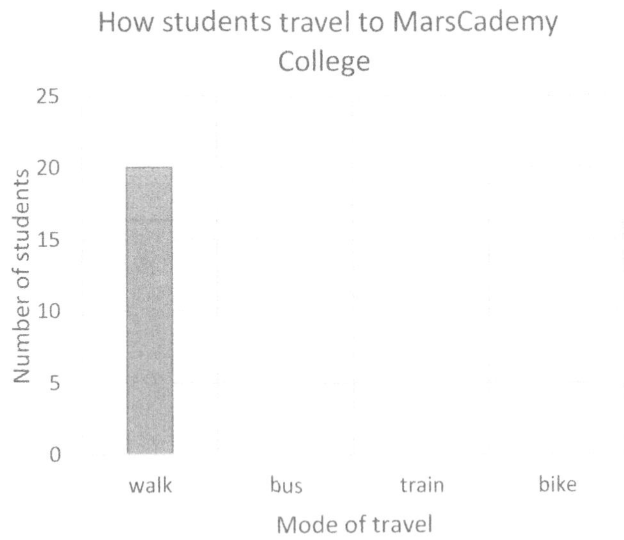

Show the missing information in the bar chart.

2 Oli works in a bedroom furniture shop. He is checking the stock levels for different items.

beds	18
wardrobes	15
bedside tables	19
dressing tables	12
mirrors	17

Show this information in a bar chart.

3 Sammy is showing her staff the number of items they have sold each month for the last six months. Show this information in the table.

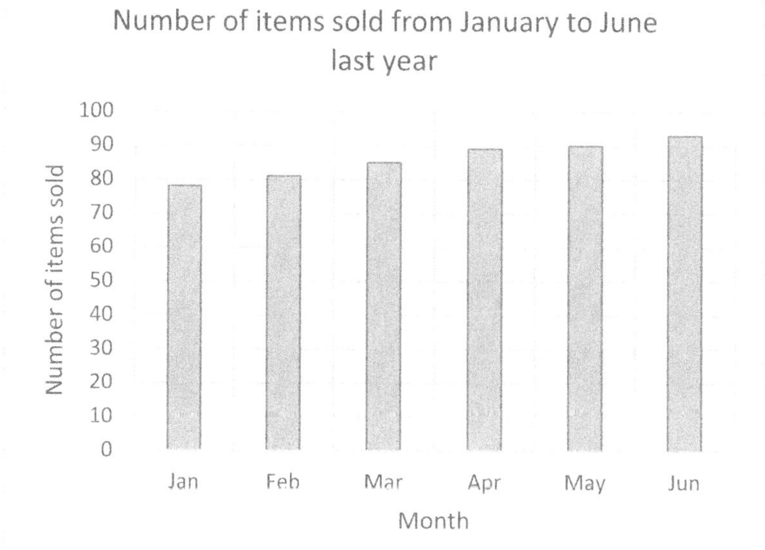

Month	**Number of items sold**

4 Find three mistakes with this bar chart.

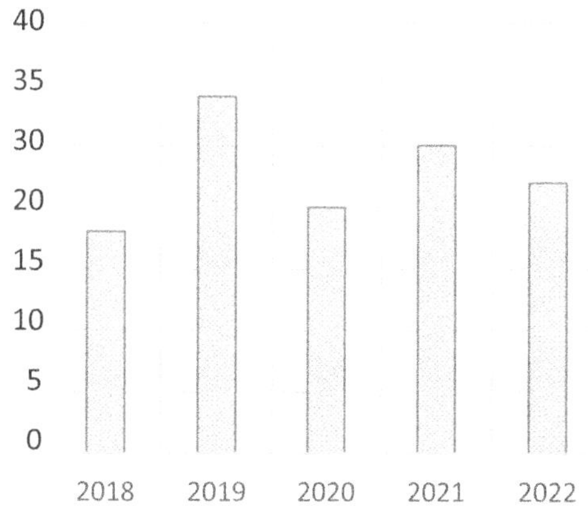

1. ______________________________

2. ______________________________

3. ______________________________

5 This table shows Jane's savings account balance over six months.

Month	Savings in £s
July	200
August	450
September	700
October	900
November	800
December	1000

Show this information in a suitable graph.

6 Jay's weekly expenditure is shown in a table.

Expense	Amount in £s
Rent	400
Groceries	200
Entertainment	150
Transport	100

Show this information in a suitable graph.

A line graph is suitable for showing data over time: over days, weeks, months or years.
Data suitable for a line graph is also suitable for a bar chart.
Data suitable for a bar chart is not always suitable for a line graph.

Progress Check

1 Rachel sees this information about recycled cups at events.

Event	Number of recycled cups
Lakeside Festival	200
Tottan Flower Show	450
Meerbrook Festival	625
South East Car Show	975

Show this information in a suitable graph.

2 Claire sees this table on a hotel website. It shows the number of guests over four years.

Year	Number of guests
2019	720
2020	350
2021	700
2022	890

Show this information in a suitable graph.

Practice Paper

Part A (Non-calculator) **20 minutes** **10 marks in total**

1 Write the number 360 in words.

______________________________(1 mark)

2 There are 215 students enrolled on a science course. 75 more students enroll on the course. How many students are enrolled on the course in total?

______________________________(2 marks)

3 Jake is buying a set of books for £45.49.
He pays with a £50 note. How much change will Jake get?

______________________________(2 marks)

4 Convert 2.45 metres into centimetres.

______________________________(1 mark)

5 Oscar is going to a football match at 15:30. He thinks this is 3:30 pm. Is he correct? Why?

______________________________(1 mark)

6 Arrange the following numbers in descending order: 625, 548, 789, 301

________, ________, ________, ________ (1 mark)

7 Tick the shape with exactly four lines of symmetry.

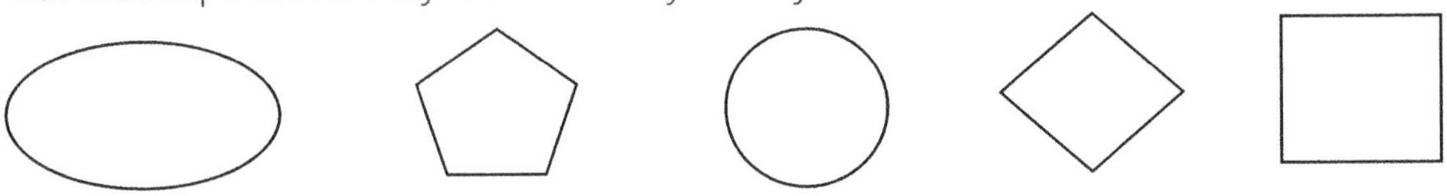

(1 mark)

8 In which genre was the highest number of books sold? ______________________________

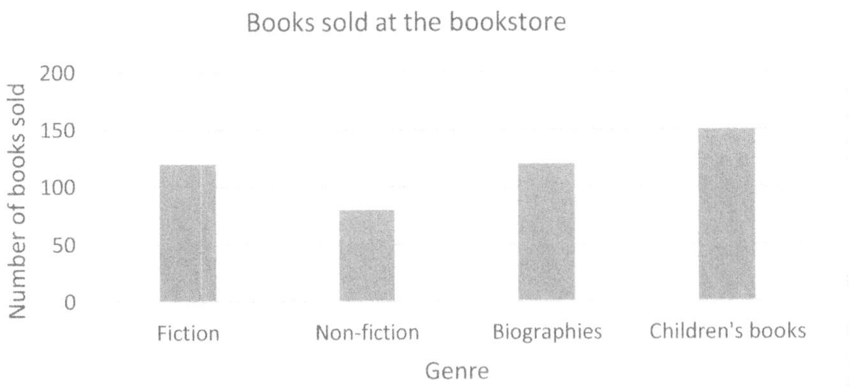

(1 mark)

Part B (Calculator)  **60 minutes** 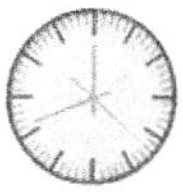 **30 marks in total**

1 Round 748 to the nearest ten.

______________________________ (1 mark)

2 Write these numbers in order, starting with the smallest: 320, 215, 400, 295

________, ________, ________, ________ (1 mark)

3 Tick the shape with the highest number of right angles.

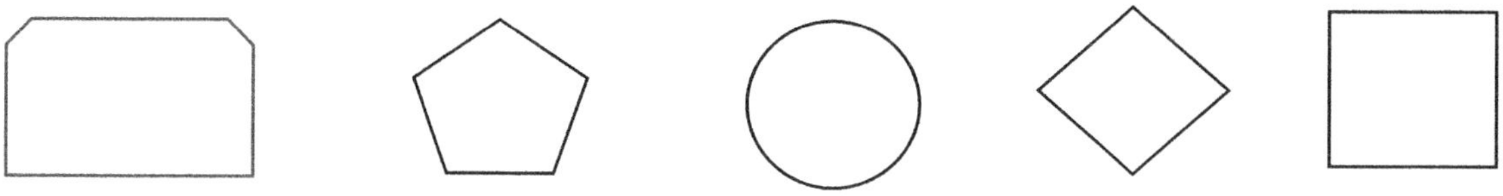

(1 mark)

4 Jack is weighing four parcels. This is how much they weigh:

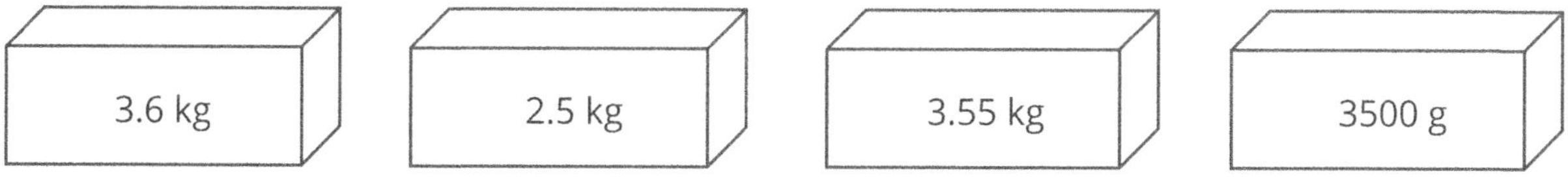

He sends the parcel that weighs the most first. Tick the parcel Jack will send first.

______________________________ (1 mark)

5 Helen is buying a new sofa. The sofa costs £480. Helen will pay a third of the cost on the first day and the rest of the amount later.

a) How much will Helen pay on the first day?

______________________________ (2 marks)

b) How much will Helen pay later?

______________________________ (1 mark)

c) Helen will pay the remainder in eight instalments. How much will each instalment be?

______________________________ (1 mark)

6 Layla is reading an 88-page book. She has already read a quarter of the pages. How many pages has she read?

______________________________ (1 mark)

Check your answer. Show your check.

______________________________ (1 mark)

7 What number comes next in the sequence: 7, 11, 15, ...?

______________________________ (1 mark)

8 The local library has provided data on the number of books borrowed over four years.

Year	Number of books
2019	450
2020	520
2021	610
2022	760

a) Show this information in a suitable graph.

(3 marks)

b) How many more books were borrowed in 2022 than 2021?

____(1 mark)

c) In which year were more than 700 books borrowed?

____(1 mark)

9 A train schedule shows your train is due to arrive at 14:45. Your current time is 13:35. How long do you have until the train arrives?

____________________ (1 mark)

10 You buy a loaf of bread (£1.50), two croissants (£0.75 each), and a pie (£2.25). You pay with a £10 note.

a) How much change will you receive?

____________________ (2 marks)

b) The bakery is currently serving ticket number 68, and your ticket number is 72. How many customers are ahead of you?

____________________ (1 mark)

11 Neil is helping to organize a community art project where participants choose their preferred paint colour. The choices recorded are as follows: Red, Blue, Red, Green, Blue, Red, Yellow, Green, Blue, Red, Yellow, Green, Blue, Red, Green, Yellow, Blue, Yellow, Red.

Colour	**Frequency**

Complete the frequency table to tally how many participants chose each colour.

(3 marks)

12 Divide 945 by 13 and express any remainder.

____________________ (1 mark)

13 Zoe thinks this jug has 60ml of liquid to the nearest labelled division. Is she correct? Explain your answer.

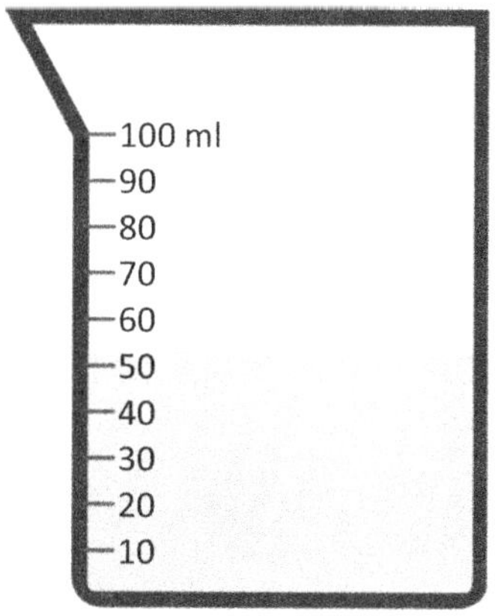

____________________ (2 marks)

14 If you are facing North and turn a half turn, which direction are you now facing?

____________________ (1 mark)

15 a) Mo and his three friends split a dinner bill of £76 equally. How much does each person pay?

____(1 mark)

b) Use estimation to check your answer.

____(1 mark)

16 Cinema tickets cost £18.55 each. How much is this amount to the nearest 10p.

____(1 mark)

Answers

Count, read and write numbers up to 1000

Q1. Page 6

101	**102**	**103**	**104**	**105**	**106**	**107**	**108**	**109**	110
111	**112**	**113**	**114**	**115**	**116**	**117**	**118**	**119**	**120**
121	**122**	**123**	**124**	**125**	**126**	**127**	**128**	**129**	**130**
131	**132**	**133**	**134**	**135**	**136**	**137**	**138**	**139**	**140**
141	**142**	**143**	**144**	**145**	**146**	**147**	**148**	**149**	**150**
151	**152**	**153**	**154**	**155**	**156**	**157**	**158**	**159**	**160**
161	**162**	**163**	**164**	**165**	**166**	**167**	**168**	**169**	**170**
171	**172**	**173**	**174**	**175**	**176**	**177**	**178**	**179**	**180**
181	**182**	**183**	**184**	**185**	**186**	**187**	**188**	**189**	**190**
191	**192**	**193**	**194**	**195**	**196**	**197**	**198**	**199**	200

Q2. Page 6

301	**302**	**303**	**304**	**305**	**306**	**307**	**308**	**309**	310
311	**312**	**313**	**314**	**315**	**316**	**317**	**318**	**319**	**320**
321	**322**	**323**	**324**	**325**	**326**	**327**	**328**	**329**	**330**
331	**332**	**333**	**334**	**335**	**336**	**337**	**338**	**339**	**340**
341	**342**	**343**	**344**	**345**	**346**	**347**	**348**	**349**	**350**
351	**352**	**353**	**354**	**355**	**356**	**357**	**358**	**359**	**360**
361	**362**	**363**	**364**	**365**	**366**	**367**	**368**	**369**	**370**
371	**372**	**373**	**374**	**375**	**376**	**377**	**378**	**379**	**380**
381	**382**	**383**	**384**	**385**	**386**	**387**	**388**	**389**	**390**
391	**392**	**393**	**394**	**395**	**396**	**397**	**398**	**399**	400

Q3. Page 6

703	**704**	705	**706**	707	**708**	709	**710**	711	**712**
450	**451**	452	**453**	454	**455**	456	**457**	458	**459**
642	643	**644**	645	**646**	647	**648**	649	**650**	651
885	886	**887**	888	**889**	**890**	**891**	892	**893**	894
225	**226**	227	**228**	229	**230**	231	**232**	**233**	**234**
549	550	**551**	552	**553**	**554**	**555**	556	**557**	**558**
888	889	**890**	**891**	**892**	893	**894**	**895**	**896**	**897**
165	**166**	167	**168**	169	**170**	**171**	**172**	**173**	**174**
561	**562**	**563**	**564**	**565**	**566**	**567**	**568**	**569**	570
307	**308**	**309**	**310**	**311**	**312**	**313**	**314**	**315**	316
792	**793**	**794**	**795**	**796**	797	**798**	**799**	**800**	**801**

Q4. Page 7

1p, 3p, 5p, 7p, 9p, 11p, 13p, 15p, 17p, 19p

Q5. Page 7

80, 82, **84**, 86, **88**, 90, **92**, 94, **96**, 98

forty seven, **forty nine**, fifty one, fifty three, **fifty five**, fifty seven, fifty nine, sixty one, sixty three, sixty five

two hundred and eighty one, two hundred and eighty three, two hundred and eighty five**, two hundred and eighty seven, two hundred and eighty nine, two hundred and ninety one,** two hundred and ninety three**, two hundred and ninety five,** two hundred and ninety seven, **two hundred and ninety nine**

572, **574,** 576, **578,** 580, **582, 584, 586,** 588, **590**

eight hundred and fifty two, eight hundred and fifty four, **eight hundred and fifty six,** eight hundred and fifty eight, **eight hundred and sixty, eight hundred and sixty two,** eight hundred and sixty four, **eight hundred and sixty six, eight hundred and sixty eight, eight hundred and seventy**

Q6. Page 7

112, 114, **116**, 118, **120**, 122, **124**
230, **232**, 234, 236, **238**, **240**
442, **444**, 445, 446, **448**, **450**
751, 753, **755**, 757, **759**, **761**, **763**
565, **567**, **569**, **571**, 573, **575**
372, **374**, **376**, 378, **380**
688, **690**, **692**, 694, **696**, **698**
890, **892**, **894**, **896**, **898**, 900

Q7. Page 7

330, 331, **332,** 333, **334,** 335**, 336**
750, **752**, 754, **756, 758**, 760
688, 686, 684, 682, **680,** 678
275, **277, 279,** 281, 283, **285**
542, **544,** 546, **548, 550, 552**
468, **470, 472,** 474, **476, 478**

Q8. Page 8

73 odd	9 **odd**	92 **even**	35 **odd**	14 **even**	516 **even**
452 **even**	45 **odd**	52 **even**	98 **even**	37 **odd**	20 **even**
648 **even**	857 **odd**	210 **even**	649 **odd**	698 **even**	243 **odd**

Q9. Page 8

Example: 202
Example: 587

Q10. Page 9

10, **20,** 30, **40,** 50, **60, 70**

Q11. Page 9

60, 70, 80, **90,** 100, **110,** 120, **130,** 140, **150**

thirty nine, forty nine, **fifty nine, sixty nine,** seventy nine, **eighty nine, ninety nine, one hundred and nine, one hundred and nineteen, one hundred and twenty nine**

153, 163, 173, **183,** 193, **203, 213, 223,** 233, **243**

two hundred and sixty five, two hundred and seventy five, two hundred and eighty five, **two hundred and ninety five, three hundred and five, three hundred and fifteen,** three hundred and twenty five**, three hundred and thirty five,** three hundred and forty five, **three hundred and fifty five**

572, **582,** 592, **602,** 612, **622, 632, 642,** 652, **662**

eight hundred and forty four, eight hundred and fifty four, **eight hundred and sixty four,** eight hundred and seventy four, **eight hundred and eighty four,**

eight hundred and ninety four, nine hundred and four, **nine hundred and fourteen, nine hundred and twenty four, nine hundred and thirty four**

Q12. Page 9
245, 255, 265, 275, 285, 295, 305 So £305.

Q13. Page 9
152, 162, 172, 182, 192, 202, 212, 222, 232 So eight notes.

Q14. Page 10
100p, 200p, 300p, **400p,** 500p, **600p, 700p, 800p, 900p**

Q15. Page 10
80, **180,** 280, **380,** 480, **580,** 680, **780,** 880, **980**

forty nine, one hundred and forty nine**, two hundred and forty nine, three hundred and forty nine,** four hundred and forty nine, **five hundred and forty nine, six hundred and forty nine, seven hundred and forty nine, eight hundred and forty nine, nine hundred and forty nine**

63, 163, 263, **363,** 463, **563,** 663, **763,** 863, **963**

eighty five, one hundred and eighty five, two hundred and eighty five, **three hundred and eighty five, four hundred and eighty five, five hundred and eighty five,** six hundred and eighty five, **seven hundred and eighty five,** eight hundred and eighty five, **nine hundred and eighty five**

Q16. Page 10
237, 337, 437, 537, 637, 737, 837, 937. So, seven notes.

Q17. Page 10
€790

Q18. Page 11
a) 85kg
b) 193kg

Q19. Page 11
450, 550, 650, 750. So, 750.

Q20. Page 11
a) Yes, she is correct if you start counting from a ten.
b) Not if you don't start from a ten.

Q21. Page 11
a) Yes, she is correct if you start counting from a hundred.
b) Not if you don't start from a hundred.

Q22. Page 11
No, 264 is an even number.

Q23. Page 11
Yes, 138 is even.

Q24. Page 12
odd + odd = even
even + even = even
odd + even = odd
even + odd = odd

Progress check Page 12

Q1.
245, **246,** 247, 248, **249, 250**

Q2.
110, 112, 114, 116, 118, **120**

Q3.
350, 400, 450, 500, 550, 600, 650. So, £650.

Q4.
690, 700, 710, 720, 730, 740, 750

Q5.
250, 130, 103, 100, 52, 0

Q6.
No, should be 180, 190, 200, 210

Read and write numbers up to 1000

Q1. Page 15

Hundreds	Tens	Units
	6	8
1	5	0
	3	**9**
4	**7**	**0**
4	**0**	**2**
7	**5**	**8**
3	**0**	**0**
4	**0**	**7**
8	**3**	**6**
9	**9**	**9**

Q2. Page 15
one hundred and eleven **111**
one hundred and twenty **120**
one hundred and sixty **160**
two hundred and seventy five **275**
four hundred and thirty four **434**
nine hundred and eighty six **986**
eight hundred and eighty six **886**
ninety nine **99**
seven hundred and seventy one **771**
three hundred and eighteen **318**
one hundred and twenty four **124**
five hundred and seven **507**
two hundred **200**
six hundred and eighty six **686**

Q3. Page 15
835

Q4. Page 16
£538 **five hundred and thirty eight** pounds
£400 **four hundred** pounds
£844 eight hundred and forty four pounds
£609 six hundred and nine pounds
£970 nine hundred and seventy pounds

Q5. Page 16
402 White Street

Q6. Page 16
859

Q7. Page 16
209

Q8. Page 17
67 **sixty seven**
113 **one hundred and thirteen**
120 **one hundred and twenty**
646 **six hundred and forty six**
258 **two hundred and fifty eight**
819 **eight hundred and nineteen**
790 **seven hundred and ninety**
366 **three hundred and sixty six**
184 **one hundred and eighty four**
500 **five hundred**
407 **four hundred and seven**
938 **nine hundred and thirty eight**

Q9. Page 17
Jasi is number **four hundred and eighty six** in the queue.

Q10. Page 17
The amount in words does not match the amount in digits.

Q11. Page 18
a) three hundred and twenty six pounds
b) yes, 379 is greater than 326.

Q12. Page 18
£628

Progress Check Page 18
Q1.

Hundreds	Tens	Units
	6	8
1	5	0
1	**4**	**1**
2	**0**	**8**
2	**9**	**0**

Q2.
117 **one hundred and seventeen**
890 **eight hundred and ninety**
601 **six hundred and one**
1000 **one thousand**

Q3.
No, this is one hundred and eighteen.

Q4.
five hundred and fifty five

Q5.

The Bank
Pay Joe Bloggs
five hundred and twenty pounds only £ 520

Order and compare numbers up to 1000

Q1. Page 19
442 is greater than 400
674 is greater than 670
370 is greater than 307
780 is less than 870

Q2. Page 19

319 is greater than 300	780 is greater than 708	560 is less than 650
560 is greater than 530	459 is less than 549	1000 is greater than 0
628 is less than 682	1000 is greater than 999	640 is greater than 365
417 is greater than 407	779 is less than 799	349 is less than 358

Q3. Page 19

429 is higher than 329	800 is higher than 776	886 is lower than 894
515 is lower than 550	654 is higher than 456	335 is lower than 366
108 is lower than 180	659 is higher than 655	95 is lower than 876
670 is lower than 760	769 is lower than 770	755 is higher than 366

Q4. Page 20
a) example: 300
b) example: 455
c) example: 311

Q5. Page 20
Out of 830, 880 and 900, 830 is the smallest and 900 is the greatest.
Out of 670, 530 and 1000, 530 is the smallest and 1000 is the greatest.
Out of 400, 150 and 999, 150 is the smallest and 999 is the greatest.

Q6. Page 20

a) 789	f) 7
b) 420	g) 189
c) 830	h) 83
d) 478	i) 298
e) 937	j) 265

Q7. Page 21

367, 391, 392	40, 140, 400
275, 280, 295	108, 180, 818
600, 620, 672	137, 145, 169

Q8. Page 21

684, 657, 603	187, 174, 153
496, 200, 163	386, 384, 353
879, 840, 804	971, 970, 969

Q9. Page 21
478m, 380m, 187m, 170m

Q10. Page 21

a) ticket C

b) ticket C, ticket A, ticket B, ticket D

Q11. Page 21

Box C

Q12. Page 21

Example: 794

Q13. Page 21

a) £489

b) £489, £499, £509

Progress Check Page 22

Q1. £981, £980, £908, £189

Q2. a)

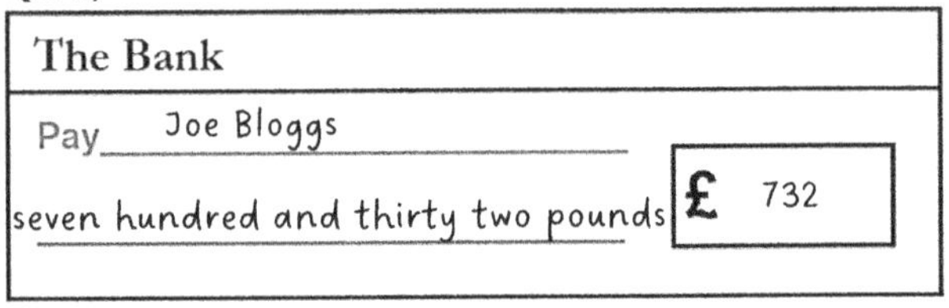

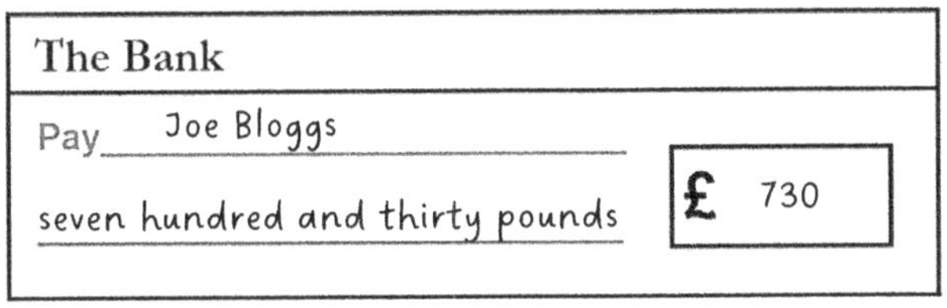

b)

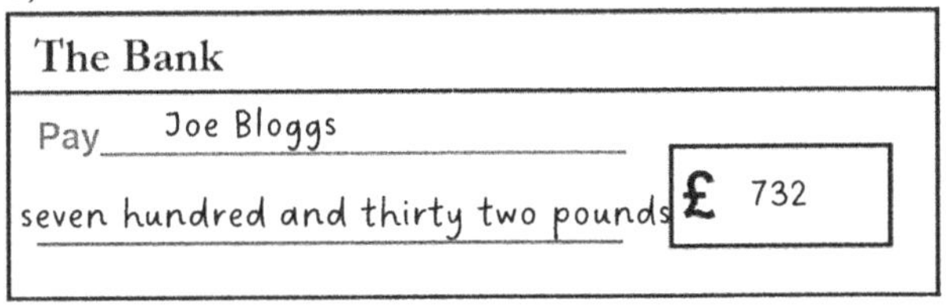

Q3.

No, should be 518, 572, 591, 603, 705.

Q4.

The Fridges

Q5.

Yes, 480 is greater than 418.

<u>Recognise and continue linear sequences of numbers up to 100</u>

Q1. Page 23

6, 12, 18, **24, 30**

Q2. Page 23

a) 4th, 7th , 10th, 13th , 16th, 19th

b) No, she will visit her mother on the 19th and then 22nd.

Q3. Page 23

a) 10p, 20p, 30p, 40p, 50p, 60p

b) Ten coins as the pattern goes: 10p, 20p, 30p, 40p, 50p, 60p, 70p, 80p, 90p, 100p.

Q4. Page 24

90p, 70p, **50p, 30p, 10p**

Q5. Page 24

a) 54, 49, 44, 39, 34

b) five

c) four

Q6. Page 24

2	4	6	**8**	**10**	**12**	**14**	**16**	**18**	**20**	**22**	**24**
5	10	15	**20**	**25**	**30**	**35**	**40**	**45**	**50**	**55**	**60**
7	**14**	**21**	**28**	35	42	49	**56**	**63**	**70**	**77**	**84**
4	**8**	**12**	**16**	**20**	**24**	**28**	32	36	40	**44**	**48**
9	**18**	27	36	45	**54**	**63**	**72**	**81**	**90**	**99**	**108**
6	**12**	**18**	**24**	**30**	**36**	**42**	**48**	**54**	60	66	72
8	16	24	**32**	**40**	**48**	**56**	**64**	**72**	**80**	**88**	**96**
3	**6**	**9**	12	15	18	**21**	**24**	**27**	**30**	**33**	**36**
11	**22**	**33**	**44**	**55**	**66**	77	88	98	110	121	132

Q7. Page 24

2, 4, 6, 8, **10, 12**
37, 35, 33, **31, 29**
74, 77, 80, **83, 86**
22, 32, 42, **52, 62**
66, 55, 44, **33, 22**
64, 72, 80, **88, 96**
52, 57, 62, **67, 72**
50, 56, 62, **68, 74**
8, 16, 24, **32, 40**
0.5, 1, 1.5, **2.0, 2.5**
3.7, 4.8, 5.9, **7.0, 8.1**
8.0, 9.2, 10.4, **11.6, 12.8**

Q8. Page 24

5, 6, 7, 8, **9, 10, 11**
28, 29, 30, **31, 32,** 33
4.5, 4.7, 4.9, **5.1, 5.3**, 5.5
14, 21, 28, **35, 42,** 49
6.6, 5.5, 4.4, **3.3, 2.2,** 1.1
34, 38, 42, **46**, 50, **54, 58**
72, **74,** 76, **78**, 80, **82**
65, 62, 59, **56, 53**, 50
2.9, 2.6, 2.3, **2.0,1.7**

Q9. Page 24

31

Q10. Page 25

10p, 20p, 30p, 40p, **50p, 60p**

Q11. Page 25

49, 58, 67, 76, 85, 94. Yes, it will be.

Q12. Page 25

99

Q13. Page 25

No, it doesn't go up by the same amount each time.

Q14. Page 25
6.5, 7, 7.5, **8, 8.5, 9**

Q15. Page 25
Yes, it is correct. You just ned to be careful when you reach the number 10 with the tens or the unit.

Progress Check page 25
Q1. Add six

Q2. 42, 36, 30, 24, **18, 12**

Q3. 55, 60, 65, **70**, 75, **80**, **85**, 90

Q4. 16, 20, 24, 28, 32, 36

Q5. Yes, this is the nine times table, so 90 will be in it.

Q6. No, should be 33, 36, 39, 42 or 43, 46, 49, 52.

Q7. Yes, that is correct for linear sequences. Other types of sequences do not need to go up or down by the same amount.

<u>Add and subtract using three-digit numbers</u>

Q1. Page 26
69 + 43 **60+ 40 = 100 9 + 3 = 12 100 + 12 = 112**
75 + 82 **70 + 80 = 150 5 + 2 = 7 150 + 7 = 157**
56 + 37 **50 + 30 = 80 6 + 7 = 13 80 + 13 = 93**
92 + 29 **90 + 20 = 110 2 + 9 = 11 110 + 11 = 121**

Q2. Page 27
620 + 206 **600 + 200 = 800 20 + 0 = 20 0 + 6 = 6**
800 + 20 + 6 = 826
353 + 204 **300 + 200 = 500 50 + 0 = 50 3 + 4 = 7**
500 + 50 + 7 = 557
463 + 235 **400 + 200 = 600 60 + 30 = 90 3 + 5 = 8**
600 + 90 + 8 = 698
469 + 226 **400 + 200 = 600 60 + 20 = 80 9 + 6 = 15**
600 + 80 + 15 = 695
305 + 236 **300 + 200 = 500 0 + 30 = 30 5 + 6 = 11**
500 + 30 + 11 = 541
392 + 243 **300 + 200 = 500 90 + 40 = 130 2 + 3 = 5**
500 + 130 + 5 = 635
534 + 547 **500 + 500 = 1000 30 + 40 = 70 4 + 7 = 11 1000 + 70 + 11 = 1081**
754 + 642 **700 + 600 = 1300 50 + 40 = 90 4 + 2 = 6**
1300 + 90 + 6 = 1396
302 + 976 **300 + 900 = 1200 0 + 70 = 70 2 + 6 = 8**
1200 + 70 + 8 = 1278
651 + 543 **600 + 500 = 1100 50 + 40 = 90 1 + 3 = 4**
1100 + 90 + 4 = 1194

Q3. Page 30

191 + 67 = **258**	124 + 485 = **609**	756 + 84 = **840**	190 + 687 = **887**	193 + 773 = **966**	913 + 86 = **999**
408 + 581 = **989**	576 + 327 = **903**	415 + 372 = **787**	284 + 426 = **710**	126 + 77 = **203**	91 + 738 = **829**
641 + 348 = **989**	470 + 158 = **628**	554 + 102 = **656**	472 + 527 = **999**	765 + 174 = **939**	792 + 109 = **901**
695 + 205 = **900**	231 + 175 = **416**	362 + 276 = **638**	559 + 310 = **869**	114 + 327 = **441**	186 + 357 = **543**
642 + 285 = **927**	173 + 235 = **408**	416 + 573 = **989**	378 + **611** = 989	468 + **364** = 732	574 + **298** = 876

Q4. Page 31
216 + 88 = 304

Q5. Page 31
656 + 239 = £895

Q6. Page 31
43 + 236 = 279GB

Q7. Page 31
149 + 214 + 96 = £459. Yes, he is correct.

Q8. Page 31
Yes. Example 17 + 0 = 17

Q9. Page 31
35 + 45 = £80

Q10. Page 31
72 + 95 =

Q11. Page 32
124 + 78 + 39 = £241

Progress Check Page 32
Q1.
499 + 449 = £948

Q2.
54 + 69 = 123

Q3.

527 + 115 = **642**	120 + **415** = 535	**300** + 99 = 399
404 + 286 = **690**	658 + **99** = 757	**435** + 68 = 503

Q4.
The first sentence is correct as order doesn't matter in addition. The second sentence is incorrect as 0 + 135 = 135 .

Q5.
Example: 125 + 346

Q6. 43 + 136 = 179

Subtract three-digit numbers

Q1. Page 33

976 – 302 **900 – 300 = 600 70 – 0 = 70 6 – 2 = 4**
600 + 70 + 4 = 674
573 – 260 **500 – 200 = 300 70 – 60 = 10 3 – 0 = 3**
300 + 10 + 3 = 313
328 – 305 **300 – 300 = 0 20 – 0 = 20 8 – 5 = 3**
0 + 20 + 3 = 23
569 – 237 **500 – 200 = 300 60 – 30 = 30 9 - 7 = 2**
300 + 30 + 2 = 332
482 – 332 **400 – 300 = 100 80 – 30 = 50 2 – 2 = 0**
100 + 50 + 0 = 150
463 – 235 **400 – 200 = 200 50 – 30 = 20 13 – 5 = 8**
200 + 20 + 8 = 228

Q2. Page 34

651 – 540 **600 – 500 = 100 50 – 40 = 10 1 – 0 = 1**
100 + 10 + 1 = 111
392 – 242 **300 – 200 = 100 90 – 40 = 50 2 – 2 = 0**
100 + 50 + 0 = 150
629 – 206 **600 – 200 = 400 20 – 0 = 20 9 – 6 = 3**
400 + 20 + 3 = 423
534 – 313 **500 – 300 = 200 30 – 10 = 20 4 – 3 = 1**
200 + 20 + 1 = 221
305 – 205 **300 – 200 = 100 0 – 0 = 0 5 – 5 = 0**
100 + 0 + 0 = 100
754 – 642 **700 – 600 = 100 50 – 40 = 10 4 – 2 = 2**
100 + 10 + 2 = 112
353 – 202 **300 – 200 = 100 50 – 0 = 50 3 – 2 = 1**
100 + 50 + 1 = 151
469 – 226 **400 – 200 = 200 60 – 20 = 40 9 – 6 = 3**
200 + 40 + 3 = 243
463 – 232 **400 – 200 = 200 60 – 30 = 30 3 – 2 = 1**
200 + 30 + 1 = 231
976 – 38 **900 - 0 = 900 60 - 30 = 30 16 - 8 = 8**
900 + 30 + 8 = 938

Q3. Page 37

191	424	756	690	793	913
- 67	- 185	- 84	- 387	- 73	- 96
124	**239**	**672**	**303**	**720**	**817**
508	576	415	484	126	791
- 481	- 327	- 372	- 226	- 77	- 538
27	**249**	**43**	**258**	**49**	**253**
641	870	954	572	765	792
- 348	- 458	- 102	- 427	- 174	- 109
593	**412**	**852**	**145**	**591**	**683**
695	831	362	559	814	386
- 205	- 175	- 276	- 310	- 327	- 157
490	**666**	**86**	**249**	**487**	**229**
642	273	516	978	768	674
- 285	- 135	- 473	- **589**	- **436**	- **96**
357	**138**	**143**	389	332	578

Q4. Page 38
699 – 120 = £579

Q5. Page 38
550 – 95 – 150 ≈ £305

Q6. Page 38
205 – 180 = 25

Q7. Page 38
Yes, she is correct.

Q8. Page 38
No, should be 330 – 245 = 85

Q9. Page 38
No, should be 580 – 410 = 170

Q10. Page 38
Yes. Example 32 – 0 = 32

Q11. Page 39
239 + ___ = 414 is the same as **414 - 239**
605 + ___ = 712 is the same as **712 - 605**
248 + ___ = 519 is the same as **519 - 248**
___ + 345 =414 is the same as **414 - 345**
___ + 877 = 912 is the same as **912 - 877**
___ + 511 = 719 is the same as **719 - 511**

Q12. Page 39
330 take away 109 is written as **330 - 109**
590 take away 258 is written as **590 - 258**
600 take away 161 is written as **600 - 161**
464 take away 384 is written as **464 - 384**
Take away 362 from 458 is written as **458 - 362**
Take away 245 from 317 is written as **317 - 245**
Take away 429 from 484 is written as **484 - 429**
Take away 178 from 859 is written as **859 - 178**

Q13. Page 40
499 – 150 = £349
Check: 349 + 150 = 499

Q14. Page 40
219 + 310 = 529
Check: 529 –310 = 219

Q15. Page 40
295 + 250 = 545
Check: 545 – 250 = 295

Q16. Page 40
35 + 80 = 115 489 – 115 = 374
Check: 115 – 80 = 35 374 + 115 = 489

Q17. Page 40
135 + 22 = 156
Check: 156 – 22 = 135

Q18. Page 40

300 + 70 + 250 = 620
620 – 300 – 70 = 250

Q19. Page 40

a) 230 – 115 = 115
b) 400 – 326 = 74
c) 732 – 584 = 148
d) 960 – 495 = 465

Q20. Page 41

a) 307 – 205 = 102
b) 250 – 175 = 75
c) 392 + 392 = 784
d) 525 + 16 = 541
e) 150 – 85 = 65
f) 350 + 200 = 550
g) 800 – 550 = 250
h) 500 – 350 = 150
i) 135 + 227 = 362
j) 256 + 64 = 320
k) 200 – 120 = 80
l) 500 – 300 = 200

Q21. Page 42

118	+	256	=	374
264	+	**268**	=	532
347	+	**294**	=	641
343	=	**72**	+	**271**
267	=	**89**	+	**178**

306	-	177	=	129
623	-	**408**	=	215
609	-	**143**	=	466
318	=	**882**		**564**
127	=	**395**	-	**268**

Q22. Page 42

118	+	256	=	374
461	+	**71**	=	532
577	+	**64**	=	641
343	=	**284**	+	**59**
267	=	**156**	+	**111**

118	+	256	=	374
115	+	**417**	=	532
76	+	**565**	=	641
343	=	**198**	+	**145**
267	=	**99**	+	**168**

Q23. Page 42

250 - 250 = 0	186 – 186 = 0	567 – 567 = 0
121 – 121 = 0	205 - 205 = 0	0 - 0 = 0

Q24. Page 42

	6	8
+	2	**7**
	9	5

	2	8
+	3	5
	6	**3**

	3	4
+	4	**8**
	8	2

	2	3
+	**3**	**8**
	6	1

Q25. Page 43

	6	8
-	2	**3**
	4	5

	8	8
-	3	**9**
	4	9

	6	4
-	4	**7**
	1	7

	7	3
-	1	**4**
	6	9

Q26. Page 43

Numbers have not been aligned, with units under units and tens under tens.

Q27. Page 43

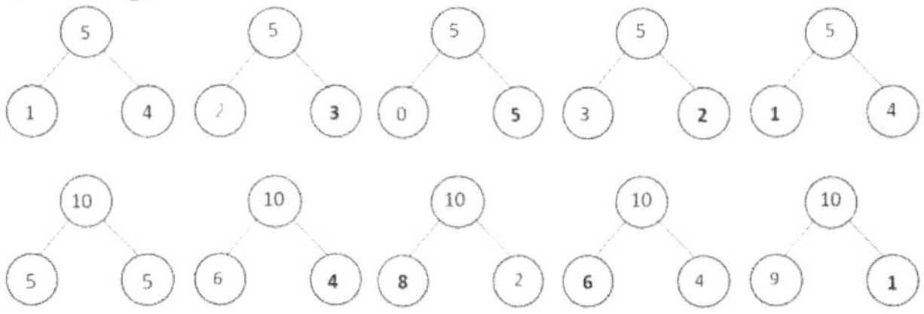

Q28. Page 44

1 + **4** = 5	2 + **3** = 5	**1** + 4 = 5	**0** + 5 = 5	**2** + 3 = 5
3 + **2** = 5	4 + **1** = 5	**3** + 2 = 5	3 + 2 = **5**	**4** + 1 = 5
0 + **10** = 10	1 + **9** = 10	**3** + 7 = 10	**10** + 0 = 10	2 + **8** = 10
3 + **7** = 10	4 + **6** = 10	**4** + 6 = 10	**8** + 2 = 10	5 + **5** = 10
6 + **4** = 10	7 + **3** = 10	**6** + 4 = 10	**1** + 9 = 10	8 + **2** = 10

Q29. Page 44

35 + 10 = 45   45 – 1 = 44
56 + 30 = 86   86 - 3 = 83
77 + 30 = 107   107 - 2 = 105
42 – 20 = 22   22 + 2 = 24
77 – 10 = 67   67 + 1 = 68
56 – 30 = 26   26 + 1 = 27

Q30. Page 45

5 + 9 + 5 =

5	+	5	=	10
10	+	9	=	19

11 + 7 + 9 =

11	+	9	=	20
20	+	7	=	27

25 - 6 - 5 =

25	-	5	=	20
20	-	6	=	14

38 - 3 - 8 =

38	-	8	=	30
30	-	3	=	27

Progress Check Page 45

Q1.
499 + 449 = 948
Yes, he is.

Q2. 112 – 86 = 25

Q3.

439 - 215 = **224**	755 - **143** = 612	**958** - 459 = 499
544 - 326 = **218**	876 - **608** = 268	**805** - 4 = 801

Q4.
The first sentence is not correct as order matters in subtraction. The second sentence is correct as subtracting 0 from a number, gives us the number itself.

<u>Multiply two-digit whole numbers by single and double-digit whole numbers</u>

Q1. Page 46

6 x 10 = **60**	11 x 10 = **110**	75 x 10 = **750**	10 x 10 = **100**	9 x 10 = **90**
7 x 10 = **70**	12 x 10 = **120**	81 x 10 = **810**	43 x 10 = **430**	10 x 89 = **890**
8 x 10 = **80**	13 x 10 = **130**	10 x 48 = **480**	10 x 37 = **370**	26 x 10 = **260**

Q2. Page 46
14 x 10 = 140

Q3. Page 46
10 x 45 = 450p

Q4. Page 46
10 x 15 = £150

Q5. Page 47

6 x 17 = **102**	9 x 15 = **135**	11 x 9 = **99**	13 x 4 = **52**	14 x 8 = **112**
7 x 13 = **91**	5 x 16 = **80**	14 x 7 = **98**	15 x 5 = **75**	16 x 9 = **144**
8 x 12 = **96**	6 x 15 = **90**	17 x 8 = **136**	11 x 8 = **88**	15 x 7 = **105**

Q6. Page 47
4 x 18 = 72p

Q7. Page 47
15 x 6 = 90

Q8. Page 47

6 x 40 = **240**	9 x 20 = **180**	9 x 80 = **720**	40 x 50 = **2000**	60 x 80 = **4800**
7 x 60 = **420**	70 x 5 = **350**	30 x 20 = **600**	70 x 70 = **4900**	30 x 80 = **2400**
8 x 30 = **240**	3 x 50 = **150**	60 x 40 = **2400**	80 x 50 = **4000**	80 x 30 = **2400**

Q9. Page 48

6 x 37 = **222**	6 x 85 = **510**	55 x 5 = **275**	9 x 35 = **315**	43 x 4 = **172**
7 x 53 = **371**	81 x 9 = **729**	21 x 8 = **168**	5 x 46 = **230**	96 x 9 = **864**
8 x 62 = **496**	74 x 7 = **518**	84 x 8 = **672**	57 x 8 = **456**	85 x 7 = **595**

Q10. Page 48
6 x 85 = 510g

Q11. Page 48
6 x 18 = £108

Q12. Page 48
7 x 54 = 378p

Q13. Page 48
6 x 60 = £360

Q14. Page 50

37 x 14

x	**30**	**7**
10	300	70
4	120	28

300 + 120 + 70 + 28 = 518

28 x 24

x	**20**	**8**
20	400	160
4	80	32

400 + 160 + 80 + 32 = 672

44 x 21

x	**40**	**4**
20	800	80
1	40	4

800 + 80 + 40 + 4 = 924

39 x 23

x	**30**	**9**
20	600	180
3	90	27

600 + 180 + 90 + 27 = 897

54 x 15

x	**10**	**5**
50	500	250
4	40	20

500 + 250 + 40 + 20 = 810

38 x 16

x	**10**	**6**
30	300	180
8	80	48

300 + 180 + 80 + 48 = 608

52 x 13

x	**50**	**2**
10	500	20
3	150	6

500 + 150 + 20 + 6 = 676

45 x 22

x	**40**	**5**
20	800	100
2	80	10

800 + 100 + 80 + 10 = 990

62 x 15

x	**10**	**5**
60	600	300
2	20	10

600 + 300 + 20 + 10 = 930

13 x 72

x	**70**	**2**
10	700	20
3	210	6

700 + 210 + 20 + 6 = 936

Q15. Page 52

4 3	7 2	2 7	5 6
x 8	x 9	x 5	x 6
2 4 4	**6 4 8**	**1 3 5**	**3 3 6**

8 1	5 7	9 2	7 3
x 4	x 3	x 5	x 6
3 2 4	**1 7 1**	**4 6 0**	**4 3 8**

6 4	8 9	6 3	9 4
x 7	x 4	x 3	x 8
4 4 8	**3 5 6**	**1 8 9**	**7 5 2**

7 2	3 8	6 2	6 7
x 6	x 5	x 8	x 9
4 3 2	**1 9 0**	**4 7 6**	**6 0 3**

Q16. Page 54

2 8	3 6	1 9	2 5
x 1 6	x 2 7	x 3 5	x 3 6
1 6 8	**2 5 2**	**9 5**	**1 5 0**
+ **2 8 0**	+ **7 2 0**	+ **5 7 0**	+ **7 5 0**
4 4 8	**9 7 2**	**6 6 5**	**9 0 0**

3 9	3 2	2 8	4 5
x 2 5	x 2 7	x 2 5	x 1 3
1 9 5	**2 2 4**	**1 4 0**	**1 3 5**
+ **7 8 0**	+ **6 4 0**	+ **5 6 0**	+ **4 5 0**
9 7 5	**8 6 4**	**7 1 0**	**5 8 5**

3 6	2 7	3 6	4 5
x 1 2	x 2 2	x 2 8	x 1 9
7 2	**5 4**	**2 8 8**	**4 0 5**
+ **3 6 0**	+ **5 4 0**	+ **7 2 0**	+ **4 5 0**
3 3 2	**5 9 4**	**1 0 0 8**	**8 5 5**

Q17. Page 55

a) 4 x 75 = £300
b) 16 x 40 = 640
c) 24 x 39 = £936
d) 9 x 32 = 288
e) 12 x 78 = £936
f) 12 x 27 = £324
g) 21 x 5 = 105
h) 24 x 40 = 960

Q18. Page 55

3 x **5** = 15 6 x **4** = 24	8 x **5** = 40 9 x **6** = 54	7 x **5** = 35 4 x **9** = 36

4 x 11 = 44 **12** x 2 = 24	**7** x 3 = 21 **6** x 10 = 60	**9** x 5 = 45 **7** x 7 = 49

Q19. Page 55

5 x 6 is the same as 6 x **5**
2 x 10 is the same as 10 x **2**
11 x 3 is the same as **3** x 11
12 x 4 is the same as 4 x **12**
7 x 8 is the same as 8 x **7**

Q20. Page 56

8 x 16 = 128

Q21. Page 56

12 x 21 = 252

Q22. Page 56

12 x 37 = £444

Q23. Page 56

Yes, as 4 has been broken down into 2 and 2.

Q24. Page 56

Yes, any number multiplied by 0 is equal to 0.

Progress Check Page 56

Q1.

26 x 4 = 104

Q2.

12 x 68 = 816

Q3.

9 x 36 = **324**   53 x 10 = **530**
6 x **12** = 72   8 x **8** = 64

Q4.

The first sentence is correct as order doesn't matter when multiplying.
The second sentence is wrong as 1 x 25 = 25.
The third sentence is wrong as 35 x 0 = 0.

Q5.

9 x 72

Divide three-digit whole numbers by single and double-digit whole numbers and express remainders

Q1. Page 58

10 ÷ 2 = **5**	40 ÷ 4 = **10**	24 ÷ 6 = **4**	42 ÷ 6 = **7**	63 ÷ 7 = **9**
15 ÷ 5 = **3**	32 ÷ 8 = **4**	16 ÷ 4 = **4**	64 ÷ 8 = **8**	100 ÷ 10 = **10**
21 ÷ 3 = **7**	16 ÷ 2 = **8**	18 ÷ 3 = **6**	81 ÷ 9 = **9**	40 ÷ 5 = **8**

Q2. Page 62

566 ÷ 2 = **283**	792 ÷ 3 = **264**	956 ÷ 4 = **239**	914 ÷ 2 = **457**	795 ÷ 5 = 159	840 ÷ 6 = 140
966 ÷ 7 = **138**	858 ÷ 3 = **286**	680 ÷ 4 = 170	873 ÷ 3 = 291	903 ÷ 7 = 129	856 ÷ 8 = 107
486 ÷ 3 = **162**	909 ÷ 9 = **101**	675 ÷ 5 = 135	486 ÷ 3 = **162**	568 ÷ 4 = 142	444 ÷ 6 = 74

Q3. Page 62

a) 360 ÷ 4 = £90
b) 498 ÷ 3 = £166
c) 165 ÷ 5 = £33
d) 920 ÷ 4 = 230
e) 144 ÷ 8 = 18
f) 375 ÷ 5 = 75

Q4. Page 62

72 ÷ 3 = 24 She didn't carry over the 1 remaining to make 12.

Q5. Page 65

569 ÷ 2 = **284rem1**	799 ÷ 3 = **266rem1**	958 ÷ 4 = **239rem2**	915 ÷ 2 = **457rem1**	798 ÷ 5 = **159rem3**	845 ÷ 6 = **140rem5**
969 ÷ 7 = **138rem3**	860 ÷ 3 = **286rem2**	683 ÷ 4 = **170rem3**	875 ÷ 3 = **291rem2**	908 ÷ 7 = **129rem5**	862 ÷ 8 = **107rem6**
487 ÷ 3 = **162rem1**	914 ÷ 9 = **101rem5**	677 ÷ 5 = **135rem2**	589 ÷ 3 = **196rem1**	567 ÷ 4 = **141rem3**	443 ÷ 6 = **73rem5**

Q6. Page 65

a) 215 ÷ 7 = 30 rem 1
b) 257 ÷ 5 = 51 rem 2
c) 794 ÷ 5 = 158 rem 4
d) 196 ÷ 9 = 21 rem 7
e) 350 ÷ 15 = 23 rem 5
f) 178 ÷ 8 = 22 rem 2

Q7. Page 65

474 ÷ 3 = She didn't carry over the 1 remaining to make 12.

Q8. Page 68

566 ÷ 12 = **47rem2**	788 ÷ 13 = **60rem8**	938 ÷ 24 = **39rem2**	905 : 22 = **41rem3**	780 ÷ 15 = **53rem5**	840 ÷ 16 = **52rem8**
954 ÷ 27 = **35rem9**	858 ÷ 23 = **37rem7**	680 ÷ 14 = **48rem**	873 ÷ 13 = **67rem2**	903 ÷ 17 = **53rem2**	825 ÷ 48 = **17rem9**
486 ÷ 33 = **14r24**	909 ÷ 19 = **47r16**	695 ÷ 25 = **27r20**	586 ÷ 53 = **11r3**	568 ÷ 24 = **23r16**	444 ÷ 26 = **17r2**

Q9. Page 68

a) 215 ÷ 7 = 30 rem 1
b) 257 ÷ 5 = 51 rem
c) 165 ÷ 5 = £33
d) 920 ÷ 4 = 230
e) 144 ÷ 8 = 18
f) 375 ÷ 5 = 75

Q10. Page 68

474 ÷ 4 = 118 rem 2, not 118.2.

Q11. Page 68

566 ÷ 12 = **47r2**	792 ÷ 13 = **60r12**	956 ÷ 24 = **39r20**	914 ÷ 22 = **41r12**	801 ÷ 15 = **53r6**	840 ÷ 16 = **52r8**
966 ÷ 27 = **35r21**	858 ÷ 23 = **37r7**	680 ÷ 14 = **48r8**	873 ÷ 13 = **67r2**	903 ÷ 17 = **53r2**	856 ÷ 48 = **17r40**
486 ÷ 33 = **14r24**	909 ÷ 19 = **47r16**	696 ÷ 25 = **27r21**	586 ÷ 53 = **11r3**	568 ÷ 24 = **23r16**	444 ÷ 26 = **17r2**

Q12. Page 68

a) 600 ÷ 15 = £40
b) 750 ÷ 12 = £6.50
c) 795 ÷ 24 = 33.125 So 33 packs.
d) 240 ÷ 14 = 17.14 So 17 pages.
e) 150 ÷ 12 = 12.5 So 12 trays.
f) 120 ÷ 7 = 17.14 So 17 pies and 1 remaining.

Q13. Page 68

366 ÷ 4 = 91.5, not 91 rem 5.

Q14. Page 70

a) 250 ÷ 5 = 50g
b) 810 ÷ 6 = £135
c) 435 ÷ 15 = 29
d) 650 ÷ 12 = 54 rem 2
e) 700 ÷ 30 = 23 rem 10
f) 630 ÷ 15 = 42
g) 882 ÷ 18 = 49
h) 760 ÷ 8 = 54 rem 4

Q15. Page 70

Yes, she converted pounds into pence.

Q16. Page 70

No, should be 975 ÷ 25 = 39

Q17. Page 70

Yes, she is correct.

Q18. Page 71

No, a number divided by 0 is undefined.

Q19. Page 71

9 x ___ = 27 is the same as **27 ÷ 9**
5 x ___ = 20 is the same as **20 ÷ 5**
8 x ___ = 32 is the same as **32 ÷ 8**
___ x 5 = 35 is the same as **35 ÷ 5**
___ x 7 = 42 is the same as **42 ÷ 7**
___ x 11 = 66 is the same as **66 ÷ 11**

Q20. Page 71

3 x 4 = 12 so 12 ÷ 3 = **4** and 12 ÷ 4 = **3**

7 x 5 = 35 so 35 ÷ 5 = **7** and 35 ÷ 7 = **5**

6 x 8 = 48 so 48 ÷ 6 = **8** and 48 ÷ 8 = **6**

4 x 7 = 28 so 28 ÷ 4 = **7** and 28 ÷ 7 = **4**

9 x 3 = 27 so 27 ÷ 3 = **9** and 27 ÷ 9 = **3**

10 x 6 = 60 so 60 ÷ 10 = **6** and 60 ÷ 6 = **10**

8 x 11 = 88 so 88 ÷ 8 = **11** and 88 ÷ 11 = **8**

4 x 12 = 48 so 48 ÷ 4 = **12** and 48 ÷ 12 = **4**

Q21. Page 72

750 + 210 = £960

Check: 960 – 210 = &50

Q22. Page 72

319 – 76 = £243

Check: 243 + 76 = 319

Q23. Page 72

156 ÷ 12 = 13

Check: 13 x 12 = 156

Q24. Page 72

12 x 39 = £468

Check: 468 ÷ 12 = 39

Q25. Page 72

7 x 120 = £840

Check: 840 ÷ 7 = 120

Q26. Page 72

244 ÷ 8 = 30 rem 4

Check: 30 x 8 = 240   240 + 4 = 248

Q27. Page 73

18 + 18 = 36 Check: 36 – 18 = 18	28 - 16 = 12 Check: 12 + 16 = 28
5 x 4 = 20 Check: 20 ÷ 5 = 4	29 + 36 = 65 Check: 65 – 36 = 29
10 + 68 = 78 Check: 79 – 10 = 68	42 ÷ 7 = 6 Check: 6 x 7 = 42
56 ÷ 8 = 7 Check: 7 x 8 = 56	37 - 6 = 31 Check: 31 + 6 = 37

Q28. Page 73

Yes, both are ways to check multiplication. Division is the opposite of multiplication; multiplication is repeated addition.

Q29. Page 73

No, should be 132 ÷ 12 = 11

Progress Check Page 73

Q1.

124 ÷ 4 = 31 packs

Q2.

700 ÷ 43 = 16 rem 12

Q3.

360 ÷ 9 = **40**	500 ÷ 10 = **50**	641 ÷ 7 = **91r4**	637 ÷ **91** = 7

Q4.

The first sentence is incorrect as order matters in division.

The second sentence is incorrect as 500 ÷ 1 = 500

Q5.

60 x 7 = 420

<u>Rounding to the nearest 10</u>

Q1. Page 75

83 **80**	119 **120**	4 **0**	427 **430**	835 **840**	550 **550**
276 **280**	54 **50**	790 **790**	599 **600**	112 **110**	75 **80**
963 **960**	271 **270**	324 **320**	68 **70**	566 **570**	105 **110**

Q2. Page 75

130 + 140 + 50 = £320

Q3. Page 75

500 – 380 = £120

Q4. Page 76

50 x 20 = 1000

Q5. Page 76

40 x 10 = 400

Q6. Page 76

646 – 182 = 464	Check: 650 – 180 = 470
116 + 578 = 694	Check: 120 + 580 = 700
933 – 37 = 896	Check: 930 – 40 = 890
647 + 285 = 932	Check: 650 + 290 = 940
559 – 442 = 117	Check: 560 – 440 = 120
170 + 189 + 535 = 894	Check: 170 + 190 + 540 = 900
259 - 75 = 184	Check: 260 – 80 = 180

Q7. Page 76

40 ÷ 8 = 5

Q8. Page 76

10 x 10 = 100

Q9. Page 76

55, 56, 57, 58, 59, 60, 61, 62, 63, 64

Q10. Page 76

175, 176, 177, 178, 179, 180, 181, 182, 183, 184

Q11. Page 76

150 + 90 + 160 = £400

Q12. Page 77

83 **100**	119 **100**	4 **0**	427 **400**	835 **800**	550 **600**
276 **300**	54 **100**	790 **800**	599 **600**	112 **100**	75 **100**
963 **1000**	271 **300**	324 **300**	68 **100**	566 **600**	105 **100**

Q13. Page 77
674 – 198 = 476 Check: 700 - 200 = 500
131 + 557 = 386 Check: 100 + 600 = 700
973 – 13 = 960 Check: 1000 - 0 = 1000
640 + 255 = 895 Check: 600 + 300 = 900
554 – 482 = 72 Check: 600 - 500 = 100
140 + 189 + 545 = 874 Check: 100 + 200 + 500 = 800
705 – 129 = 576 Check: 700 - 100 = 600

Q14. Page 77
Highest possible number: **250**
Lowest possible number: **349**

Q15. Page 77
148 + 162 = £310. Rounded to nearest hundred: **£300**.

Q16. Page 78
a) £259 + £474 = **£733**
b) 260 + 470 = **£730**

Q17. Page 78
a) 12 x £18 = **£216**
b) £10 x 20 = **£200**.

Q18. Page 78
132, 157, 186 rounded to nearest 10: 130 + 160 + 190 = **£480**.

Q19. Page 78
34 km, 78 km, 59 km rounded to nearest 10 km: 30 + 80 + 60 = **170 km**.

Q20. Page 78
742, 658, 395 rounded to nearest 100: 700 - 400 = **300**.

Q21. Page 78
Estimating Cost for Students:
312 students x £12/notebook, rounded: 310 x £10 = **£3,100**.

Q22. Page 78
Decorations £123, Food £245, Music £87, rounded: £120 + £250 + £90 = **£460**.

Q23. Page 78
Rent £650, Groceries £230, Utilities £120, rounded: £700 + £200 + £100 = **£1,000**.

Q24. Page 79
823 + 275, rounded to nearest 100: 800 + 300 = **1,100**.

Q25. Page 79
265 miles + 378 miles, rounded: 300 + 400 = **700 miles**.

Q26. Page 79
Emma's rounding mistake: 637 - 258 should be 600 - 300 = **300**.

Progress Check Page 79
Q1.
a) 296 + 298 = **£594**.
b) Estimation: 300 + 300 = **£600**.

Q2.
a) 281 + 319 + 356 = **£956**.
B) Estimation: 280 + 320 + 360 = **£960**.

Q3.
a) Numbers Rounded to 60:
b) Possible Original Numbers: **55, 56, 57, 58, 59, 60, 61, 62, 63, 64**.

Q4.
Numbers Rounded to 500:
Highest Possible Number: **549**. Lowest Possible Number: **450**.

Q5.
Jade's Rounding of 30:
Jade's answer: **30**. Explanation: 30 rounded to nearest 10 is **30**.

Q6.
a) 43 to nearest 10: **40**.
b) 43 to nearest 100: **0**.
c) 430 to nearest 10: **430**.
d) 430 to nearest 100: **400**.

Read, write and understand thirds, quarters, fifths and tenths including equivalent forms

Q1. Page 81

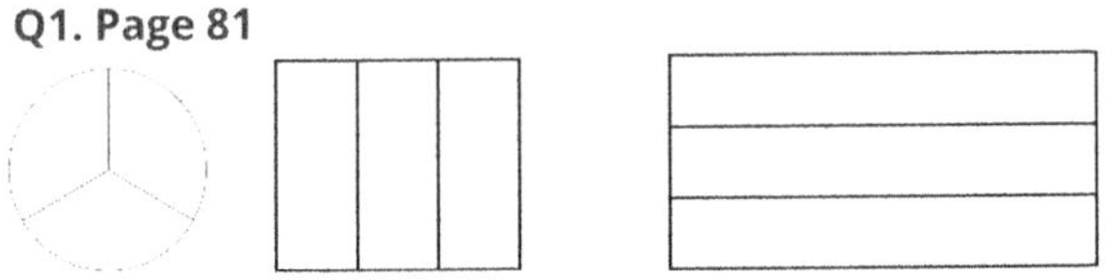

Q2. Page 81
No, the parts are not equal.

Q3. Page 81

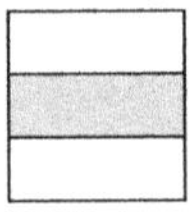

Q4. Page 81

Q5. Page 81
$\frac{1}{3} < \frac{2}{3}$ $\frac{3}{3} > \frac{2}{3}$ $\frac{3}{3} > \frac{1}{3}$

Q6. Page 81
Six cupcakes ticked.

Q7. Page 82
a) 12 ÷ 3 = 4 eggs
b) 30 ÷ 3 = £10

Q8. Page 82

$\frac{2}{6}$ or $\frac{1}{3}$

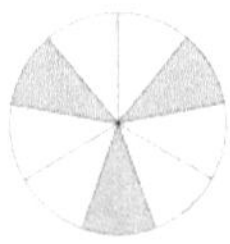

$\frac{3}{9}$ or $\frac{1}{3}$

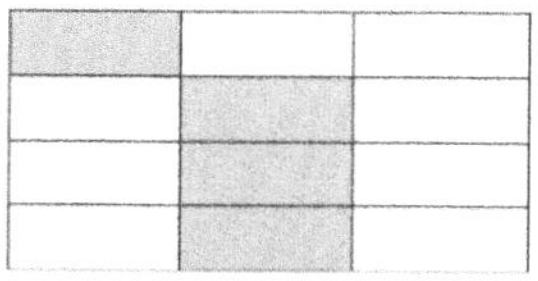

$\frac{4}{12}$ or $\frac{1}{3}$

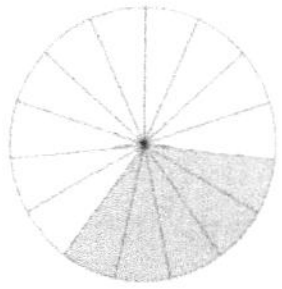

$\frac{5}{15}$ or $\frac{1}{3}$

Q9. Page 83

Q10. Page 83

Yes, all parts are equal in size.

Q11. Page 83

Q12. Page 83

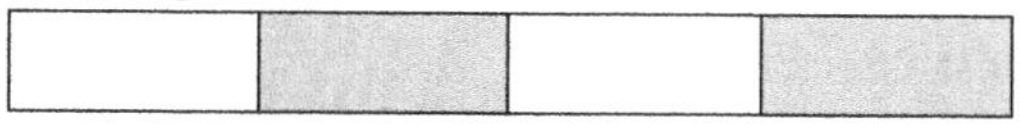

Q13. Page 83

$\frac{2}{4} < \frac{3}{4}$ $\frac{1}{4} < \frac{4}{4}$ $\frac{2}{4} > \frac{1}{4}$

Q14. Page 84

a) 8 ÷ 4 = 2 slices

b) 12 ÷ 4 = 3 sweets

c) 16 ÷ 4 = 4 cans of tuna

d) 20 ÷ 4 = £5

e) 12 ÷ 4 = 3 cherries

Q15. Page 84

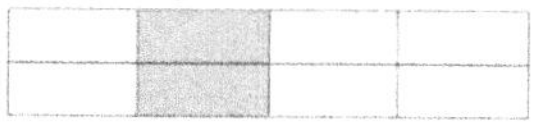

$\frac{2}{8}$ or $\frac{1}{4}$

$\frac{9}{12}$ or $\frac{3}{4}$

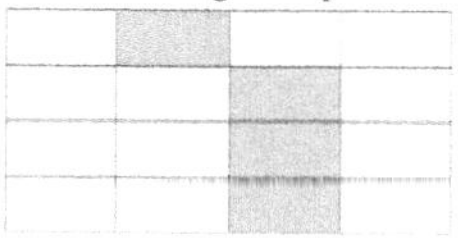

$\frac{4}{16}$ or $\frac{1}{4}$

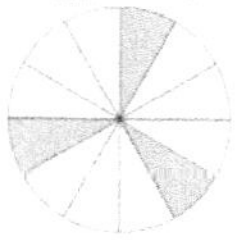

$\frac{3}{12}$ or $\frac{1}{4}$

Q16. Page 85

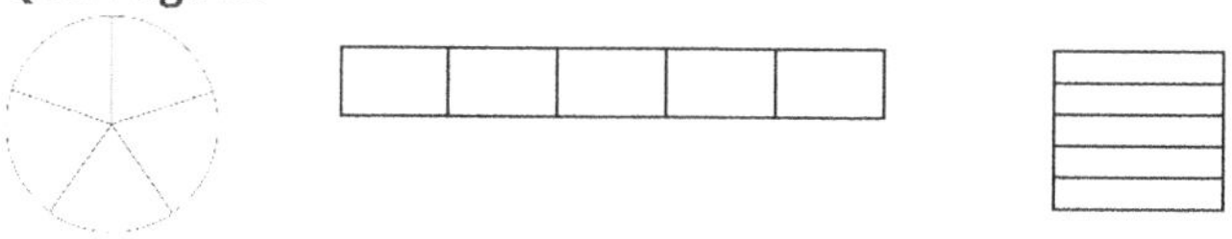

Q17. Page 85

No, the parts are not equal.

Q18. Page 85

60 ÷ 5 = 12

Q19. Page 85

$\frac{1}{5}$ is left.

Q20. Page 85

30 ÷ 5 = 6

Q21. Page 86

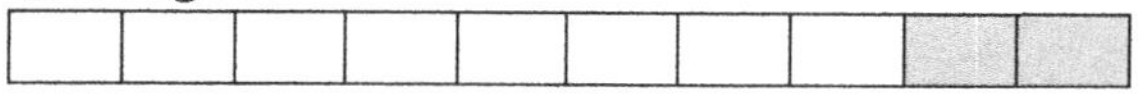

Q22. Page 86

Q23. Page 86

$\frac{3}{5} > \frac{1}{5}$ $\frac{1}{4} > \frac{1}{5}$ $\frac{5}{5} = \frac{4}{4}$

Q24. Page 86

$\frac{1}{5}$

Q25. Page 86

$\frac{2}{10}$ or $\frac{1}{5}$

$\frac{4}{10}$ or $\frac{2}{5}$

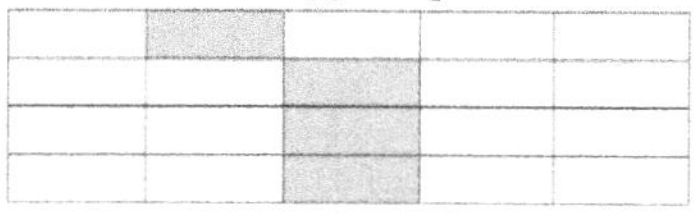

$\frac{4}{20}$ or $\frac{1}{5}$

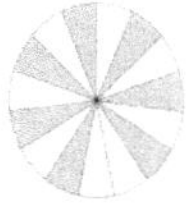

$\frac{6}{15}$ or $\frac{3}{5}$

Q26. Page 87

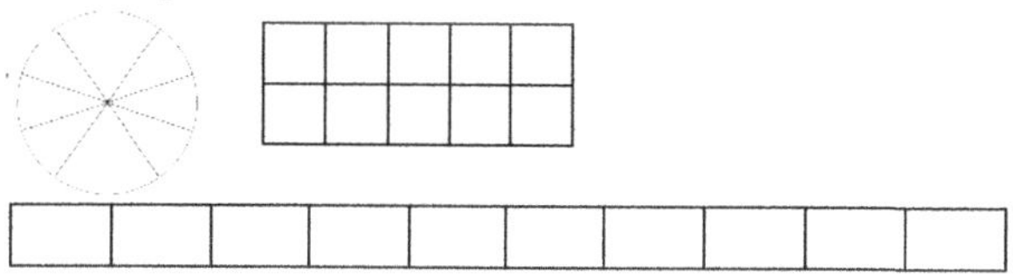

Q27. Page 87

No, the parts are not equal.

Q28. Page 87

a) 20 ÷ 10 = £2

b) 40 ÷ 10 = 4 sweets

c) 10 ÷ 10 = 1 egg

d) 120 ÷ 10 = £12

Q29. Page 88

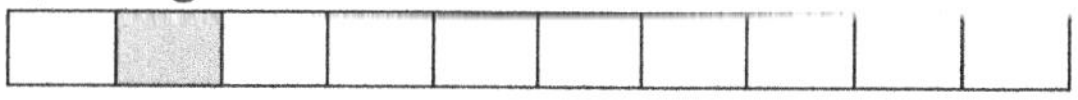

Q30. Page 88

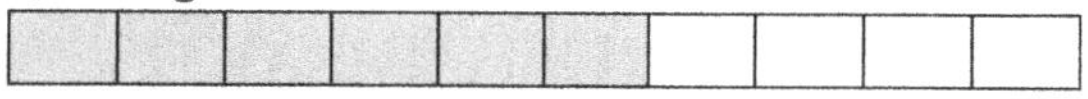

Q31. Page 88

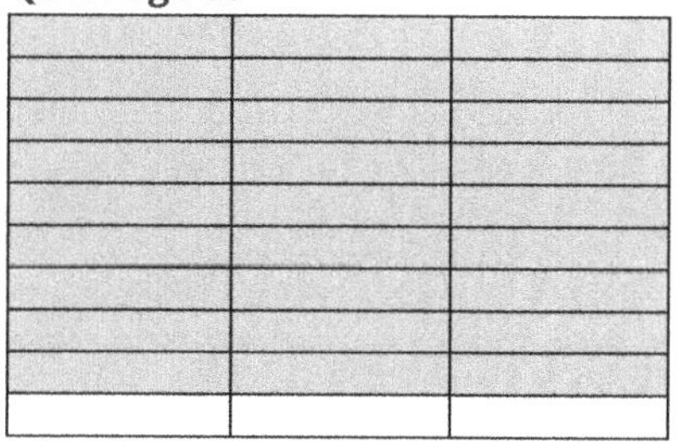

Q32. Page 88

$\frac{2}{10} < \frac{8}{10}$ $\frac{2}{5} = \frac{1}{10}$ $\frac{1}{4} > \frac{2}{10}$

Q33. Page 88

$\frac{1}{4}$ $\frac{1}{2}$ $\frac{1}{4}$ $\frac{1}{2}$

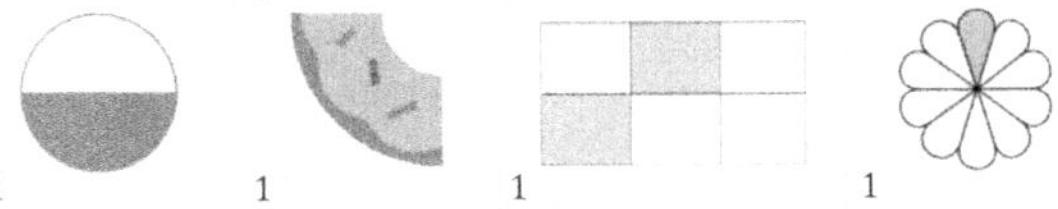

$\frac{1}{2}$ $\frac{1}{4}$ $\frac{1}{3}$ $\frac{1}{10}$

$\frac{1}{2}$ $\frac{1}{10}$ $\frac{2}{10}$ $\frac{1}{2}$

Q34. Page 89

No, 20 ÷ 4 = 5.

Q35. Page 89

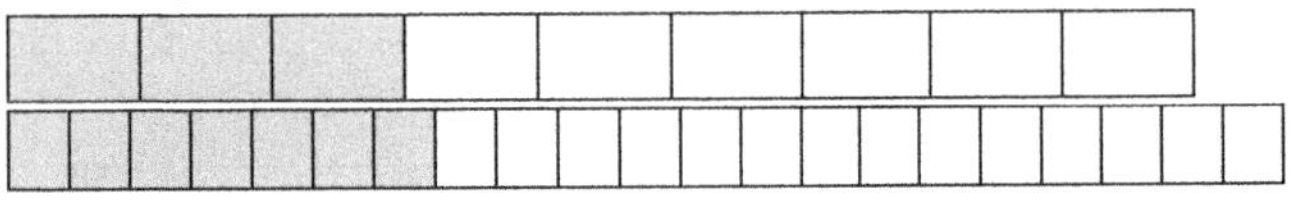

Q36. Page 89

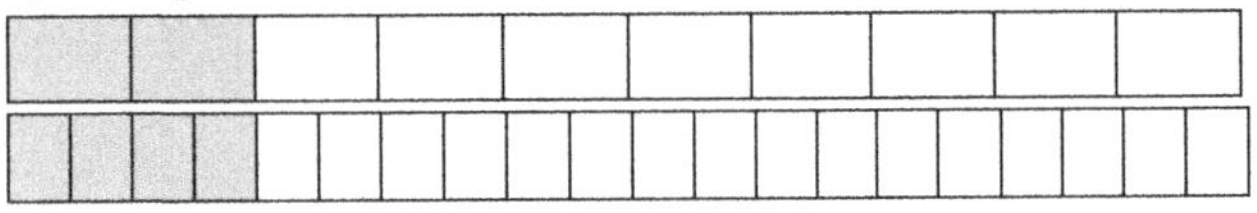

Q37. Page 89

$\frac{3}{3} - \frac{1}{3} = \frac{2}{3}$

Q38. Page 89

a) 2

b) 2

c) 3

d) 5

Q39. Page 89

a) $\frac{1}{2}$

b) $\frac{1}{2}, \frac{1}{3}, \frac{1}{4}, \frac{1}{5}, \frac{1}{10}$

c) $\frac{1}{5}$

d) $\frac{1}{5}, \frac{1}{4}, \frac{4}{10}, \frac{1}{2}, \frac{3}{3}$

Q40. Page 89

$\frac{3}{12}$ or $\frac{1}{4}$

Q41. Page 90

Yes, 8 ÷ 2 ÷ 2 = 2 and 8 ÷ 4 = 2

Q42. Page 90

No, 40 ÷ 4 = 10cm

Q43. Page 90

$\frac{9}{15}$ or $\frac{3}{5}$

Q44. Page 90

998 ÷ 2 = 499

Q45. Page 90

400 ÷ 4 = 100   400 − 100 = £300

Q46. Page 90

$\frac{10}{30}$ or $\frac{1}{3}$

Q47. Page 90

a) 500 ÷ 10 = £50

b) 500 − 50 = £450

Q48. Page 90

15 ÷ 5 = 3   15 − 3 = 12 classes remain

Q49. Page 90

12 ÷ 3 = 4m

Progress Check Page 91

Q1.

Q2.

Yes, $\frac{1}{3}$ also means 1 out of 3.

Q3.

Fraction in digits	Fraction in words
$\frac{3}{4}$	three quarters
$\frac{2}{5}$	two fifths
$\frac{6}{10}$	six tenths

Q4.

$\frac{1}{4}$ of 84 = 21

$\frac{1}{5}$ of 100 = 20

So, $\frac{1}{4}$ of 84 is greater.

Q5.

$\frac{1}{10} = \frac{7}{70}$

Q6.

No, $\frac{1}{10}$ means splitting something into ten parts, while $\frac{1}{3}$ means splitting it into three.

Q7.

120 ÷ 4 = 30g.

Read, write and used decimals up to two decimal places

Q1. Page 92

3.40 --- three point four zero

16.7 --- sixteen point seven

0.95 --- zero point nine five

10.02 --- ten point zero two

140.3 --- one hundred and forty point three

Q2. Page 94

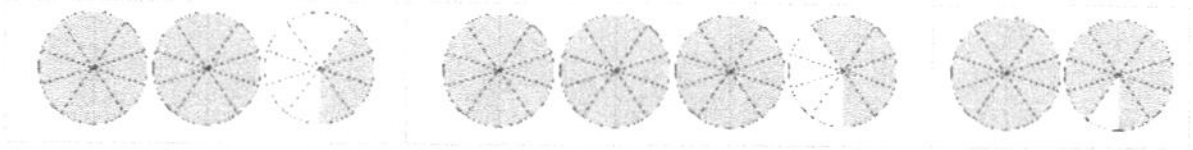

2.4 3.6 1.9

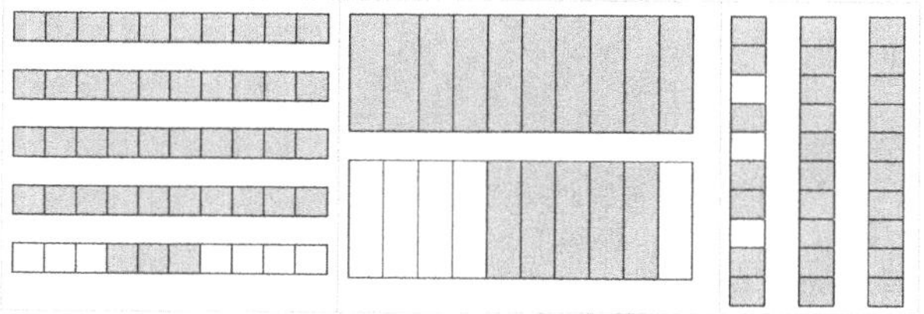

4.3 1.5 2.7

Q3. Page 94

0.31 0.75

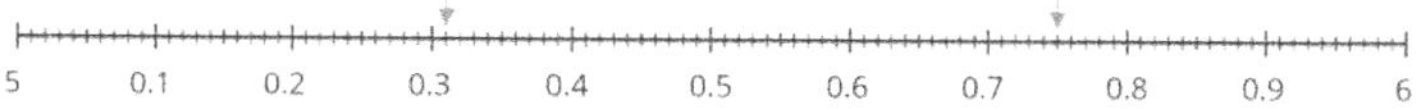

10.13 10.50

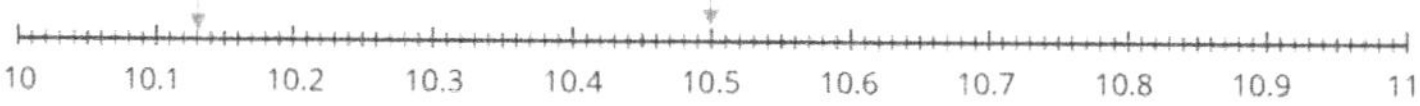

64.27 64.83

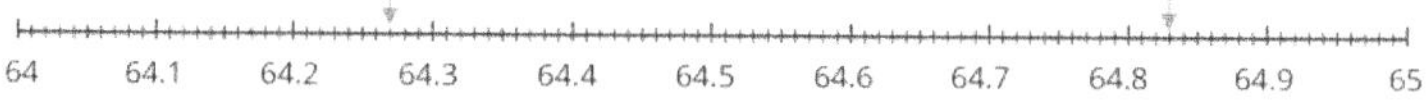

Q4. Page 95

32.15m 32.50m 32.83m

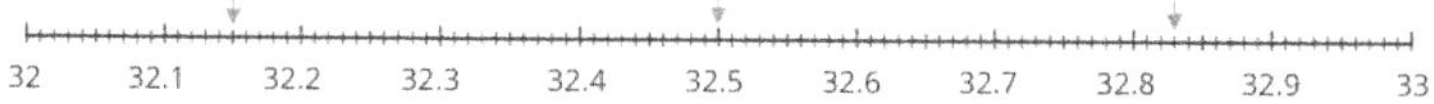

59.25km 59.43km 59.77km

20.02kg 20.41kg 20.97kg

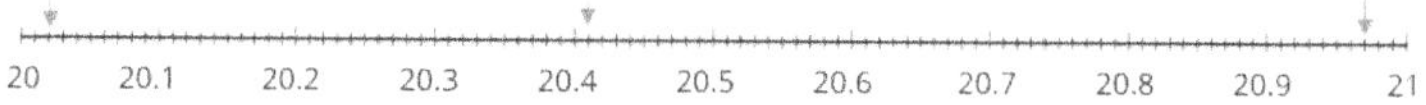

Q5. Page 95

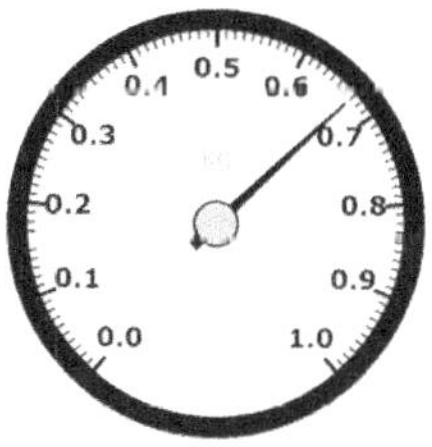

0.67

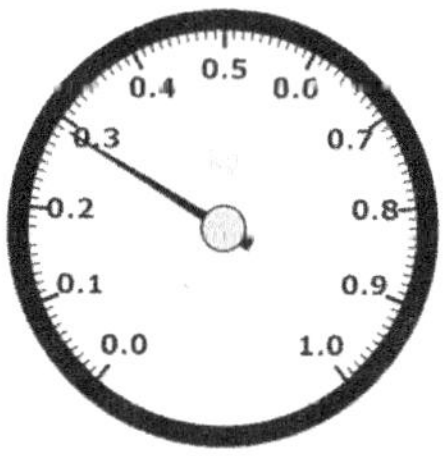

0.29

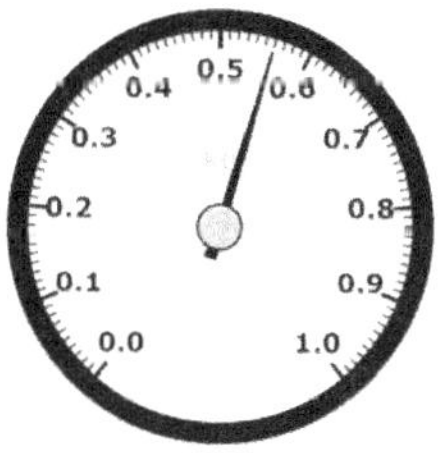

0.56

Q6. Page 95

a)tenths b) 9 c) 5.76

Q7. Page 95

0.75, 0.80, 1.08, 1.80

Q8. Page 96

6.7	5.5	5.6	5.0	7.34	11.39	1.46
+ 3.1	+ 4.2	+ 2.4	+ 3.7	+2.37	+ 8.67	+5.78
9.8	**9.7**	**8.0**	**8.7**	**9.71**	**20.06**	**7.24**
4.84	7.64	1.55	8.49	2.66	9.1	6.9
+4.17	+1.76	+7.29	+ 2.6	+7.78	+3.86	+7.49
9.01	**9.40**	**8.84**	**11.09**	**10.44**	**12.96**	**14.39**

Q9. Page 96

6.7	5.5	8.6	9.0
- 3.1	- 4.4	- 5.4	- 8.7
3.6	**1.1**	**3.2**	**0.3**
8.1	7.63	7.2	8.3
- 4.84	- 2.7	- 5.16	- 2.67
3.26	**4.93**	**2.04**	**5.63**

9.35	9.37	5.74
- 7.36	- 4.68	- 4.59
1.99	**4.69**	**1.15**
7.0	9.0	7.4
- 2.7	- 3.84	- 6.84
4.3	**5.16**	**0.56**

Q10. Page 97

a) C, B, D, A

b) 1 + 1.5 = 2.5kg

c) Friday 4.65km

d) 0.75 + 0.75 = 1.5*l*

e) A 2.30m

f) 0.75 – 0.4 = 0.35

Q11. Page 97

a) 0.6

b) 0.67

c) 1.1

d) 20.3

e) 1.0

Q12. Page 98

$\frac{1}{4}$ = 0.25

$\frac{1}{5}$ = 0.2

$\frac{1}{10}$ = 0.1

$\frac{3}{4}$ = 0.75

$\frac{1}{3}$ = 0.33

Q13. Page 98

467.14

Q14. Page 98

823.8

Q15. Page 98

12.01, 12.1, 12.11, 12.21

Q16. Page 98

11.5, 23, 34.5, **46**

Q17. Page 98

0.75 kg + 0.25 kg = **1 kg**

Progress Check Page 99

Q1.

£ 1 1 . 1 1

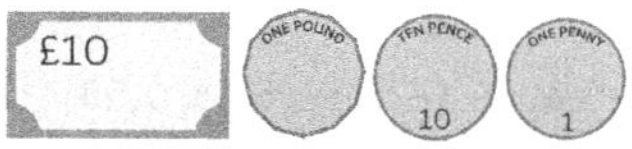

Q3.
a) 16.6 + 16.6 = 33.2cm
b) Yes, 33.2cm is longer than 30cm.

Q4.

Decimal in digits	Decimal in words
0.67	zero point six seven
14.09	**fourteen point zero nine**
400.9	four hundred point nine

Q5.
3.75 and 3.57 have the same digits but 3.75 is greater as it has 7 tenths, not five.

Q6.
a) 45.7, 47.5, 54.7, 57.4, 74.5, 75.4, 4.57, 4.75, 5.47, 5.74, 7.45, 7.54
b) 4.57, 4.75, 5.47, 5.74, 7.45, 7.54, 45.7, 47.5, 54.7, 57.4, 74.5, 75.4.

Recognise and continue sequences that involve decimals

Q1. Page 100
1.2, 2.4, 3.6, **4.8, 6.0**

Q2. Page 100
0.75, 1.5, 2.25, **3, 3.75, 4.5**

Q3. Page 100
No, should be 1.0.

Q4. Page 101
2.40, 2.20, 2.00, **1.80, 1.60**

Q5. Page 101
a) 5, 4.75, 4.5, **4.25, 4**
b) 5 - 4.75 = 0.25kg
c) 5, 4.75, 4.5, 4.25, 4, 3.75, 3.5, 3.25, 3, 2.75, 2.5, 2.25, 2, 1.75, 1.5, 1.25, 1, 0.75, 0.5, 0.25, 0. So, 20 packs.

Q6. Page 101
0.2, 0.4, 0.6, 0.8, **1.0, 1.2**
3.7, 3.5, 3.3, **3.1, 2.9**
7.4, 7.7, 8.0, **8.3, 8.6**
0.22, 0.32, 0.42, **0.52, 0.62**
1.66, 1.55, 1.44, **1.33, 1.22**
6.4, 7.2, 8.0, **8.8, 9.6**
10.52, 10.57, **10.62, 10.67**
9.50, 9.56, 9.62, **9.68, 9.74**
0.8, 1.6, 2.4, **3.2, 4.0**

Q7. Page 101
5.5, 5.6, 5.7, **5.8, 5.9, 6.0**
2.7, 2.8, 3.0, **3.1, 3.2,** 3.3
4.5, 4.7, 4.9, **5.1, 5.3,** 5.5
1.4, 2.1, 2.8, **3.5, 4.2,** 4.9
6.6, 5.5, 4.4, **3.3, 2.2,** 1.1
5.34, 5.38, 5.42, **5.46**, 5.50, **5.54, 5.58**

Q8. Page 101
6 - 0.5 = 5.5

Q9. Page 101
1.2, 1.8, 2.4, 3.6, 4.2, 4.8, 5.4, 6.0, 6.6, 7.2. No, 7.0 is not in the sequence.

Q10. Page 101
2.8 - 2.5 = 0.3 2.8 + 0.3 = 3.1

Q11. Page 101
Yes, just need to be careful when reaching 10 with the units.

Q12. Page 102
2.05m

Q13. Page 102
0.3 m, 0.5 m, 0.7 m, **0.9** m, **1.1** m

Q14. Page 102
0.4 kg, 0.8 kg, **1.2** kg, **1.6** kg

Q15. Page 102
2.5 - 2.0 = 0.5 3.0 + 0.5 = 3.5cm

Q16. Page 102
41 - 20.5 = 20.5
61.5 + 20.5 = 82
20.5, 41, 61.5, **82**

Q17. Page 102
B. 4.0

Q18. Page 102
B. 0.1

Q19. Page 102
4.0, 4.2, 4.4, **4.6**

Progress Check Page 102
Q1.
3.3, 3.4, **3.5**, 3.6, **3.7**, 3.8
6.1, 6.4, **6.7**, 7.0, **7.3**
1.2, 2.4, 3.6, **4.8**, **6.0**, **7.2**

Q2.
No, should be 2.4, 3.6, 4.8, 6.0, 7.2

Q3.
Yes, that is correct for linear sequences.

Calculate with money using decimal notation and express money correctly in pounds and pence

Q1. Page 103

hundreds	tens	units	.	tenths	hundredths
	2	1		0	0
1	3	5		0	5
4	0	5		8	9
		0		0	5

Q2. Page 103

£0.77 = 77p £0.80 = 80p £0.08 = 8p £0.70 = 70p £0.99 = 99p
£0.09 = 9p £0.90 = 90p £0.50 = 50p £0.05 = 5p

Q3. Page 104

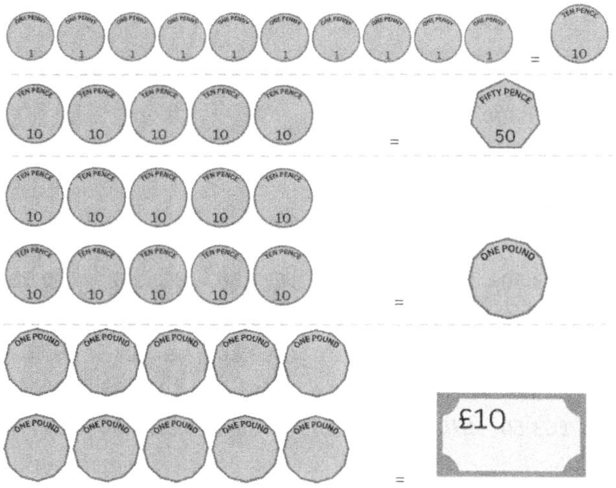

Q4. Page 104

a) 100p or £1 b) 200p or £2

Q5. Page 104

£1.77 = 177p £3.80 = 380p £4.08 = 408p £1.70 = 170p £5.99 = 599p
£5.09 = 509p £2.90 = 290p £9.50 = 950p £2.50 = 250p

Q6. Page 104

£50.78

Q7. Page 106

403.87	112	186.40	530.80	376.46	197.35	363.54
+ 339.08	+ 78.90	+ 476.54	+ 74.65	+ 533.73	+ 456.65	+ 365.45
742.95	**190.90**	**662.94**	**605.45**	**910.19**	**654.00**	**728.99**
463.80	374.66	105.36	436.44	583.72	318.26	473.64
+ 374.51	+ 46.27	+ 0.93	+ 6.57	+ 37.46	+ 372.98	+ 437.37
838.31	**420.93**	**106.29**	**443.01**	**621.18**	**691.24**	**911.01**
378.47	350.47	376.44	362.46	374.55	100	365.48
+ 475.83	+ 437.83	+ 400.00	+ 75.36	+ 47.46	+ 47.69	+ 357.67
854.30	**788.20**	**776.44**	**437.82**	**422.01**	**147.69**	**723.15**

Q8. Page 106

6.50 + 4.20 = 10.70
10.70 + 10.70 = £21.40

Q9. Page 106

23.99 + 37.99 = £61.98

Q10. Page 106

12.99 + 1.85 = £14.84

Q11. Page 106

0.39 + 0.39 = £0.78

Q12. Page 106

300.95 + 275.35 = £576.30

Q13. Page 106

15.50 – 12.75 = £2.75

Q14. Page 108

403.87	112.90	186.40	530.80	376.46	497.35	365.54
- 339.08	- 78.90	- 176.54	- 74.65	- 233.73	- 156.65	- 363.45
64.79	**34.00**	**9.86**	**456.15**	**142.73**	**340.70**	**2.09**
463.80	374.66	105.36	436.44	583.72	378.26	473.64
- 374.51	- 46.27	- 0.93	- 6.57	- 37.46	- 312.98	- 437.37
89.29	**328.39**	**104.43**	**429.87**	**546.26**	**65.28**	**36.27**
478.47	450.47	476.44	362.46	374.55	159.80	365.48
- 375.83	- 437.83	- 400.00	- 75.36	- 47.46	- 47.69	- 357.67
102.64	**12.64**	**76.44**	**287.10**	**327.09**	**112.11**	**7.81**

Q15. Page 108

135.45 + 56.30 = £191.75

Q16. Page 108

Gym A

Q17. Page 108

79.99 – 15.00 = £64.99

Q18. Page 108

142.25 + 98.00 +
200.75 = £441.00

Q19. Page 108

75 – 41.30 = £33.70

Q20. Page 109

a.i) 18.85 + 26.99 + 11.99 = £57.93
a.ii) 70 – 57.93 = £12.03
b) 27.99 + 19.99 = £47.98
£50 – 47.98 = £2.02
c) The pound should be put in front of the amount: £17
d) 39.99 + 39.99 = £79.98
e) No, it's 50 dollars, not pounds.
f) No. it's £37.
g) No, she spent £37, not 37p.
h) You can only write £15 for 15 pounds or 15p for 15 pence, not both symbols at the same time.

Q21. Page 109

a)

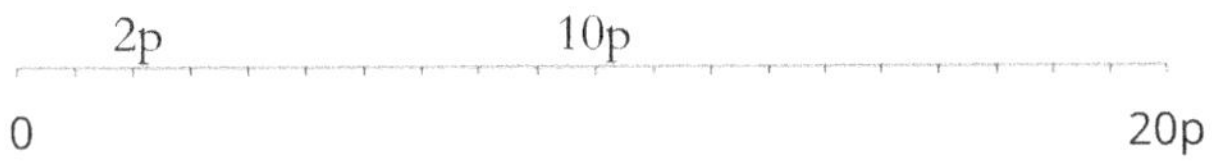

b)

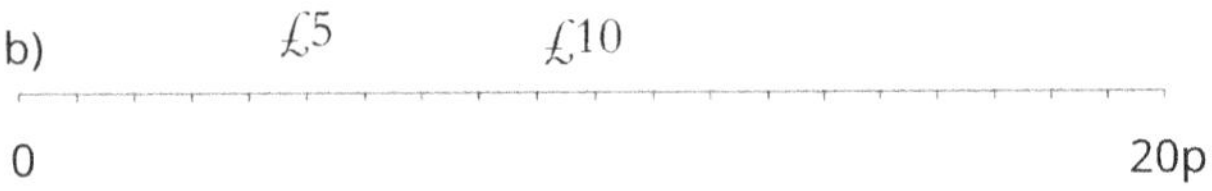

Q22. Page 109

20 – 12.95 = £7.05

Q23. Page 109

1.69 + 1.69 + 1.69 = £5.07

Progress Check Page 110

Q1.

£4.50 is worth 4 pounds and 50 pence
£4.05 is worth 4 pounds and 5 pence
£0.45 is worth 0 pounds and 45 pence

Q2.

a) 295.99 + 455.99 = £751.98
b) 1000 – 751.98 = £248.02

Q3.

£18.99 + £43.99 = **£62.98**	£66.83 – £34 = **£32.83**
£35.45 + **£64.55** = £100	£79 – **£9.30** = £69.70

Q4.

She has not aligned them correctly.
Should be:

£56.99
+ £29.52

Round amounts of money to the nearest £1 or 10p

Q1. Page 112

£83.57	£19.03	£4.36	£27.26	£35.50	£50.49
£84	**£19**	**£4**	**£27**	**£36**	**£50**
£76.00	£54.65	£90.69	£99.80	£0.36	£0.75
£76	**£55**	**£91**	**£100**	**£0**	**£1**
£63.50	£70.00	£24.10	£168.56	£376.84	£527.12
£64	**£70**	**£24**	**£169**	**£277**	**£527**

Q2. Page 112

Lowest possible amount: £6.50
Highest possible amount: £7.49

Q3. Page 112

Lowest possible amount: £22.50
Highest possible amount: £23.49

Q4. Page 112

£3

Q5. Page 114

£83.57	£19.03	£4.36	£27.26	£35.50	£50.49
£83.60	**£19.00**	**£4.40**	**£27.30**	**£35.50**	**£50.50**
£76.00	£54.65	£90.69	£99.80	£0.36	£0.75
£76.00	**£54.70**	**£90.70**	**£99.80**	**£0.40**	**£0.80**
£63.50	£70.00	£24.10	£168.56	£376.84	£527.12
£63.50	**£70.00**	**£24.10**	**£168.60**	**£376.80**	**£527.10**

Q6. Page 114

£38

Q7. Page 114

£2.80

Q8. Page 114

£77

Q9. Page 114

£789

Q10. Page 114

£28 (Shirt) + £43 (Coat) + £6 (Umbrella) = £77

Q11. Page 115

Video Game Price Rounded:
£49.99 → £50.00.

Q12. Page 115

64.75 – 18.76 = 45.99
Rounded: 65 – 19 = 46
11.50 + 57.46 = 68.96
Rounded: 12 + 57 = 69
93.99 – 3.10 = 90.89
Rounded: 94 – 3 = 91

Q13. Page 115

55.73 – 42.88 = 12.85
Rounded: 55.70 – 42.90 = 12.80
10.63 + 19.95 + 55.14 = 85.72
Rounded: 10.60 + 20.00 + 55.10 = 85.70
75.59 – 29.45 = 46.14
Rounded: 75.60 – 29.50 = 46.10

Q14. Page 115

Lowest possible amount: £7.25.
Highest possible amount: £7.34.

Q15. Page 115

Lowest possible amount: £23.55.

Highest possible amount: £23.64.

Q16. Page 115

9 x 9 = £81

Q17. Page 115

a) 60 + 40 + 60 = 160m

b) 160 ÷ 25 = 6.4 So 7 lengths are needed.

c) 7 x 30 = £210

Q18. Page 116

No, you round the figures before calculating.

So, 90 x 8 = £720

Q19. Page 116

Yes, that is correct. £0.90 = 90p

Progress Check Page 116

Q1.

£4.50 to the nearest pound is **£5**.

£4.50 to the nearest 10p is **£4.50** .

£4.05 to the nearest pound is **£4**.

£4.05 to the nearest 10p is **£4.10**.

£0.45 to the nearest pound is **£0**.

£0.45 to the nearest 10p is **£0.50**.

Q2.

a) 295.99 + 455.99 = 751.98

b) Example: 296 + 456 = 752

Q3.

a) Lowest: £5.50 Highest: £6.49

b) Lowest: £6.75 Highest: £6.84

Q4.

No, £73.99 rounded to the nearest £1 is £74.

Read time from analogue and 24-hour clocks in hours and minutes

Q1. Page 118

03 : 05 07 : 40 06 : 20 03 : 55

Q2. Page 118

04:55 06:50 05:35 09:10

10:06 06:18 02:37 04:21

Q3. Page 118

10 past 7 6 : 13 ten to 11 forty past 9

Q4. Page 119

21:12 | 19:45 | 14:30 | 07:12 | 08:55

quarter to eight twelve past nine 12 past 7 fourteen thirty 5 to 9

Q5. Page 119

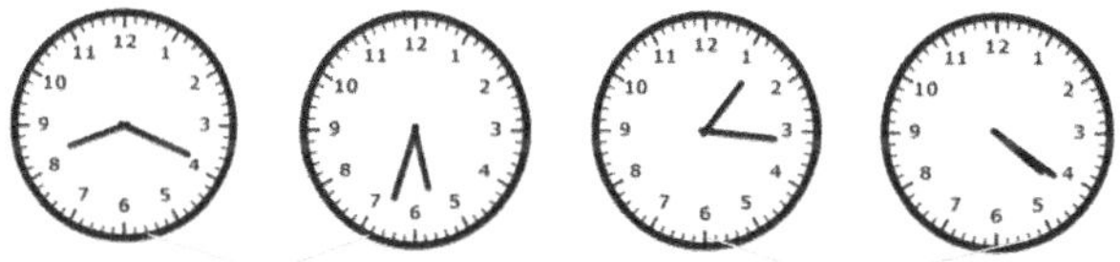

17:33 | 08:19 | 04:21 | 01:16

Q6. Page 119

Five twenty	three fifty three	ten thirty six	seven forty two
Twenty past five	seven to four	twenty four to eleven	eighteen to eight
08:35	10:06	20:42	16:50
Eight thirty five	Ten zero six	Forty two past twenty	Ten to five
Twenty five to nine	Six past ten	Eighteen to nine	Sixteen fifty

Q7. Page 119

07:12	14:30	08:55	07:08
12 past 7	fourteen thirty	5 to 9	8 past 7

Q8. Page 120

a) 09:15 or 9:15am

b) 10pm

c)

d) 10:35 + 1:45 = 12:20. No, she will not finish by 12:00.

e) 1:15pm
f) 9:02 + 3:00 = 12:02
g) Friend A
h) 9:30 + 2:30 = 12:00
i) 2:15 + 1:30 = 3:45pm Yes, he will.
j) 5:53 + 1:45 = 7:38pm

Q9. Page 121
No, this is 2:20 am. In the afternoon that would be 14:20.

Q10. Page 121
7:45 + 1:30 = 9:15am. No, he finishes 15 mins late.

Q11. Page 121
No, it's 25 minutes until 4pm. There are 60 minutes in one hour, not 100.

Progress Check Page 121

Q1.

2:20pm 4:50am 7:35pm 10:55am

Q2.

10:44 | 16:10 | 12:21 | 01:25

Q3.
Seven forty five
Forty five past seven

Q4.

a) 

b) No, this is quarter to eleven. The long hand is at 9 and the short hand at nearly 11.

Read, measure and record time using am and pm

Q1. Page 123

05:30 | 07:42 | 09:13 | 10:48 | 11:06

5:30am 7:42am 9:13am 10:48am 11:06am

Q2. Page 123

13:09 | 18:13 | 15:26 | 20:59 | 23:35

1:09pm 6:13pm 3:26pm 8:59pm 11:35pm

Q3. Page 123

02:43 06:27 09:19 10:35
2:43am 6:27am 9:19am 10:35am

Q4. Page 123

14:43 18:27 21:19 22:35
2:43pm 6:27pm 9:19pm 10:35pm

Q5. Page 123
9:00 AM + 3 hours = 12:00 PM
12:00 PM + 45 minutes = 12:45 PM

Q6. Page 123
5:00 PM - 3:30 PM = 1 hour and 30 minutes

Q7. Page 124
7:45 PM + 2 hours = 9:45 PM

Q8. Page 124
11:55 PM - 2 hours = 9:55 PM
9:55 PM - 45 minutes = 9:10 PM

Q9. Page 124
6:30 AM + 45 minutes = 7:15 AM
7:15 AM + 12 minutes = 7:27 AM

Q10. Page 124
7:30 AM to 5:00 PM = 9 hours + 30 minutes = 9.5 hours

Q11. Page 124
10:30 PM to 7:00 AM = 8 hours + 30 minutes = 8.5 hours

Q12. Page 124
7:00 PM - 1 hour = 6:00 PM
6:00 PM - 15 minutes = 5:45 PM
Bill leaves at 6:00 PM, so he is not correct.

Q13. Page 124
3:15 PM - 35 minutes = 2:40 PM

Q14. Page 124
8:15 PM + 1 hour = 9:15 PM
9:15 PM + 45 minutes = 10:00 PM

Q15. Page 124
5:20 AM + 3 hours = 8:20 AM
8:20 AM + 15 minutes = 8:35 AM

Q16. Page 124
8:30 AM + 9 hours = 5:30 PM

Q17. Page 125
8:45 AM to 5:00 PM = 8 hours + 15 minutes =
8 hours 15 mins - 1 hour lunch = 7 hours 15 mins

Progress Check Page 125

Q1.

14:20 04:50 19:35 10:55

Q2.

a) 12:30 PM - 11:12 AM = 1 hour 18 minutes

b) 20 minutes + 10 minutes + 30 minutes = 60 minutes

Q3.

a) 9:10 PM + 30 minutes = 9:40 PM

b) 10:45 AM - 2 hours = 8:45 AM

Q4.

a) 3 minutes x 12 pancakes = 36 minutes

b) 60 minutes / 3 minutes per pancake = 20 pancakes

Q5.

Enes needs to finish by 11:00 AM - 1 hour = 10:00 AM

10:00 AM - 10 minutes = 9:50 AM

He starts at 09:55, so he is not on time.

Compare metric measures of length including millimetres, centimetres, metres and kilometres

Q1. Page 126

7cm
10cm
6cm
12cm
8cm
15cm
11c

Q2. Page 127

16.2cm and 15.5cm

Q3. Page 127

70mm as this is the only one longer than 60mm.

Q4. Page 128

170mm

Q5. Page 128

549m

Q6. Page 128

59cm, 602mm, 620mm, 63.1cm

Q7. Page 128

0.1cm and 0.10cm

Q8. Page 129

0.1m = 10 cm	0.2m = **20** cm	0.3m = **30** cm	0.4m = **40** cm	0.5m = **50** cm
0.6m = **60** cm	0.7m = **70** cm	0.8m = **80** cm	0.9m = **90** cm	0.11m = **11** cm
0.25m = **25** cm	0.80m = **80** cm	**0.65** m = 65cm	**0.82** m = 82cm	**6.41** m = 641cm
9.47 m = 947cm	**1.63** m = 163cm	**3.94** m = 394cm	**3.7** m = 370cm	**4.68** m = 468cm

Q9. Page 129

1.5m

Q10. Page 129

A and D

Q11. Page 130

Yes, 86.2 + 86.2 cm=172.4

Q12. Page 130

0.83 m

Q13. Page 130

0.95 m

Q14. Page 130

191 cm.

Q15. Page 130

151 cm (1.51 m).

Q16. Page 130

2.04 m−1.25 m=0.79 m2.04 m−1.25 m=0.79 m.

Q17. Page 130

Width of room: 2.4 m = 240 cm

Width of shelf: 1.5 m = 150 cm

Space for lamp: 80 cm

Total space needed: 150 cm+80 cm=230 cm

Space available: 240 cm−230 cm=10 cm remaining.

Both fit.

Q18. Page 131

0.1km = 100m	0.2 km = **200** m	0.3 km = **300** m	0.4km = **400** m	0.5km = **500** m
0.6km = **600** m	0.7km = **700** m	0.8km = **800** m	0.9km = **900** m	0.11km = **110** m
0.25km = **250** m	0.80km = **800** m	**0.65** km = 650m	**0.82** km = 820m	**0.641** km = 641m
0.947 km = 947m	**0.163** km = 163m	**3.94** km = 3940m	**3.705** km = 3705m	**4.689** km = 4689m

Q19. Page 131

Town C, Town B, Town A, Town D

Q20. Page 131

1.3 km+0.7 km=2.0 km

Q21. Page 131

6.2 km, 7.5 km, 8.9 km, 10.1 km

Q22. Page 132

kilometre	millimetre	metre	centimetre
km	**mm**	**m**	**cm**

Q23. Page 132
millimetre, centimetre, metre, kilometre

Q24. Page 132
3.7 km × 2 = 7.4 km

Q25. Page 132
a) 3m
b) 2cm
c) 20mm
d) 80cm

Q26. Page 132
Office furthest away: B (1.65 km)

Q27. Page 133
No, 2.1 meters is 210 cm, which is greater than 200 cm.

Progress Check page 133

Q1.
6 m = 600 cm
450 cm = **4.5 m**
2.4 km = 2400 m

Q2.
a) 116 cm
b) 1.16m
c) 116 – 96 = 20cm or 0.2m

Q3.
700 mm (70 cm), **64 cm**, **0.8 m** (80 cm)

Q4.
No, the wardrobe is too long by 2 cm.

Compare measures of height including grams and kilograms

Q1. Page 134

kilogram	gram
kg	**g**

Q2. Page 134
1 kilogram

Q3. Page 134
5 grams, 50 grams, 5 kilograms, 50 kilograms

Q4. Page 135

0.1 kg = 100 g	0.2 kg = **200** g	0.3 kg = **300** g	0.4kg = **400** g	0.5kg = **500** g
0.6kg = **600** g	0.7kg = **700** g	0.8kg = **800** g	0.9kg = **900** g	0.11kg = **110** g
0.25kg = **250** g	0.80kg = **800** g	**.65** kg = 650g	**0.82** kg = 820g	**0.641** kg = 641g
0.947 kg = 947g	**0.163** kg = 163g	**3.94** kg = 3940g	**3.705** kg = 3705g	**4.689** kg = 4689g

Q5. Page 135

2kg 500g 3kg 250g 0kg 750g 2kg 750g

Q6. Page 136
19.8kg, 20.9kg, 21.0kg, 22.5kg

Q7. Page 136
8 stone
0.6kg
16g

Q8. Page 136
1.5kg and 1900g

Q9. Page 136
1000 – 650 = 350g

Q10. Page 136
No, 17.8 + 82.5 = 100.3kg

Q11. Page 137
a) Circle: **10kg**
b) Circle: **100g**
c) Circle: **1g**
d) Circle: **200g**

Q12. Page 137
Sara's answer about the TV unit weighing 20g is incorrect because 20g is too light; TVs are much heavier.

Q13. Page 137
Charles is correct; **10.0kg** is the same as **10kg**.

Q14. Page 137
3×30g+7g=97g. Yes, Paul can send the letter because it's under 100g.

Q15. Page 137
No, it's 65g.

Q16. Page 137
30 kg, **3200 g** (3.2 kg) and **4800 g** (4.8 kg)

Q17. Page 137
800 g jar.

Q18. Page 137
5.19 kg

Q19. Page 138
Jimmy's statement is correct. The package of rice weighs 1500g which is heavier than a 1 kg (1000g) bag of flour.

Q20. Page 138
Sarah is not correct. 2500 grams is equal to 2.5 kilograms, which is heavier than 2.4 kilograms.

Q21. Page 138
Yes, 4kg + 300g = 4.3kg

Progress Check Page 138
Q1.
4kg = **4000 g**
3500g = **3.5 kg**

Q2.
a) 450g = 0.45kg
b) 0.5kg = 500g
c) 500 – 450 = 50g or 0.05kg

Q3.
400g, **0.9kg** (900g), **6.4kg**, **6500g**

Q4.
No, the potatoes weigh 7kg.

Compare measures of capacity including millilitres and litres

Q1. Page 139

litre	millilitre
l	***ml***

Q2. Page 139
1 litre

Q3. Page 139
5 millilitres, 50 millilitres, 5 litres, 50 litres

Q4. Page 140

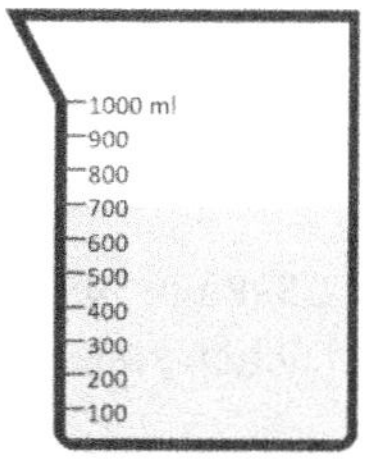

70 ml

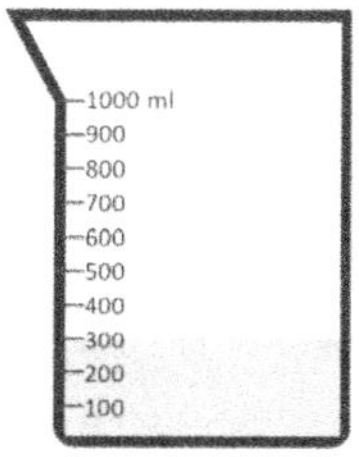

30 ml

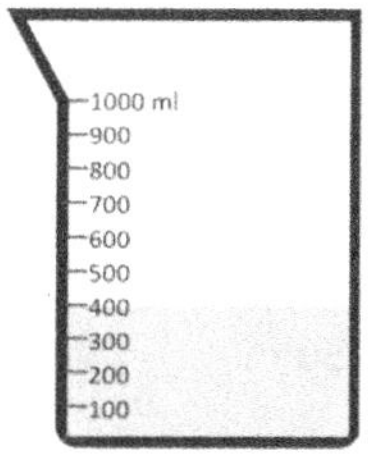

40 ml

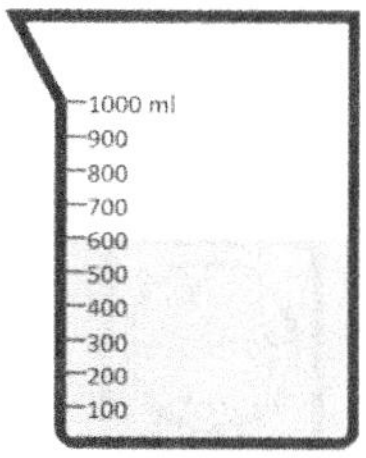

60 ml

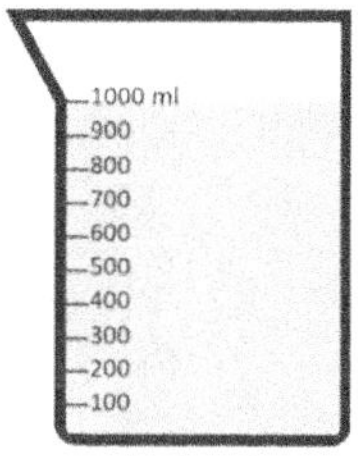

100 ml

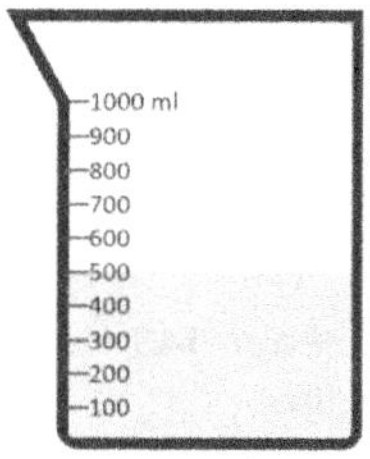

50 ml

Q5. Page 140
15 ÷ 5 = 3

Q6. Page 140
12 ÷ 1.5 = 8

Q7. Page 140
1.5 ÷ 0.4 = 3.75 So, 4 cans of soup.

Q8. Page 141
a) 20 + 70 = 90ml
b) $\frac{1}{2}$
c) 12 x 1 = 12 litres
d) 0.5 + 0.5 + 0.5 + 0.5 = 2 litres. So, 4 bottles.
e) No, it holds more than 100ml.
f) 15 ÷ 2 = 7.5 So 7 bottles

Q9. Page 141
Yes, 12.0 is equal to 12.

Q10. Page 142
a) 1 litre b) 100 millilitres c) 10 litres

Q11. Page 142

0.1 l = 100 ml	0.2 l = **200** ml	0.3 l = **300** ml	0.4 l = **400** ml	0.5 l = **500** ml
0.6 l = **600** ml	0.7 l = **700** ml	0.8 l = **800** ml	0.9 l = **900** ml	0.11 l = **110** ml
0.25 l = **250** ml	0.80 l = **800** ml	**0.65** l = 650 ml	**0.82** l = 820 ml	**0.641** l = 641 ml
0.947 l = 947 ml	**0.163** l = 163 ml	**3.94** l = 3940 ml	**3.705** l = 3705 ml	**4.689** l = 4689 ml

Q12. Page 142
70 + 93 = 163 ml

Q13. Page 143
70 litres, 7 litres, 70 millilitres, 7 millilitres

Q14. Page 143
Yes, 8 x 12 = 96 litres

Q15. Page 143
Should be 100 litres.

Q16. Page 143
400 – 350 = 50ml

Q17. Page 143
C - 650 ml

Q18. Page 143
a) 1000ml and 1litre
b) $\frac{1}{2}$ litre

Q19. Page 143
D 85 ml

Q20. Page 143
1.05 litre

Q21. Page 144
1000 ml

Q22. Page 144
Jasmine's coffee cup that holds the most coffee: **550 ml**
The assistant should give the customer the **2 litre** carton.

Q23. Page 144
2 litres

Progress Check Page 144
Q1.
7l = 7000 ml
5500ml = 5.5 l

Q2.
a) 0.7*l*
b) 300ml
c) 700 – 300 = 400ml or 0.4*l*

Q3.
0.45l (450ml), **990ml**, **4300ml**, **4.5L** (4500ml)

Q4.
No, there is 700ml. The mark is halfway between 600 and 800ml.

Use a suitable instrument to measure mass and length

Q1. Page 145
kitchen scale, body scale

Q2. Page 145
measuring tape

Q3. Page 145
kitchen scale

Q4. Page 145
measuring tape

Q5. Page 145
kitchen scale

Q6. Page 145
measuring tape

Q7. Page 146

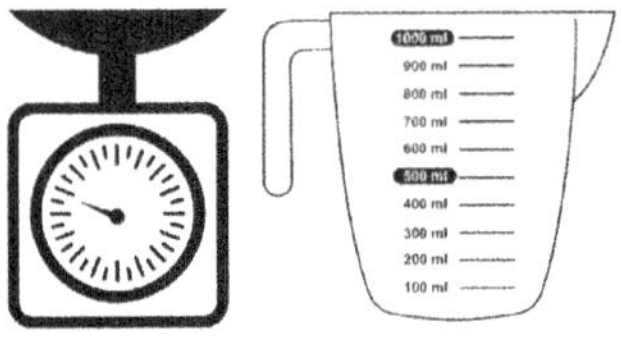

Q8. Page 146

Q9. Page 146
measuring jug

Q10. Page 146
body scale

Progress Check Page 146
Q1.

measuring tape kitchen scale ruler body scale measuring wheel

Q2.
Measuring tape or ruler.

Q3.
Kitchen scale for food items or body scale for larger items like luggage.

Use and compare measures length, weight, capacity and temperature using metric or imperial units to the nearest labelled or unlabelled division

Q1. Page 147

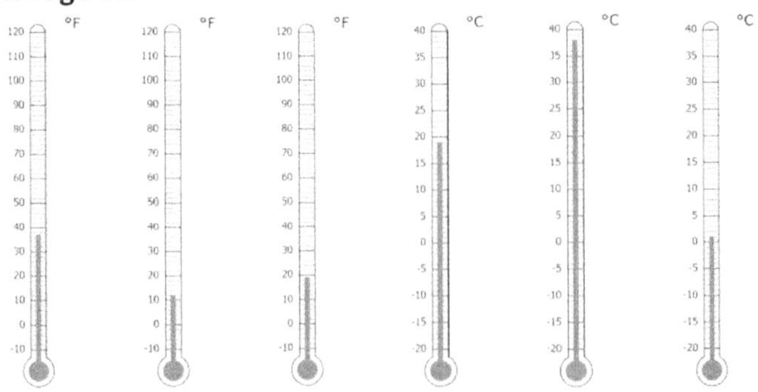

40°F 10 °F 20 °F 20 °C 40°C 0 °C

Q2. Page 148

a)

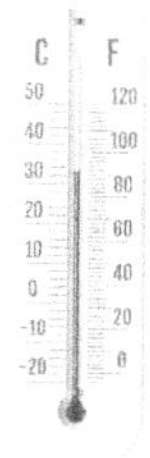

b) 30°C   80°F

Q3. Page 148

Yes, it's half-way between 26 and 27, so to the nearest cm it is 27.

Q4. Page 148

26cm

17cm

Q5. Page 148

No, it is 26cm.

Q6. Page 148

88.5, 88.6. 88.7, 88.8. 88.9, 89.0, 89.1, 89.2, 89.3, 89.4

Q7. Page 149

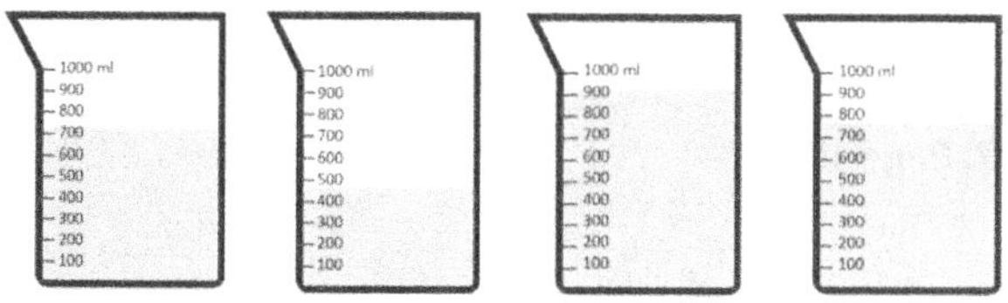

700ml   500ml   900ml   800ml

Q8. Page 149

120g   200g   100g   140g

Q9. Page 149

a) 80kg   b) No, it's 0 litres

Q10. Page 150

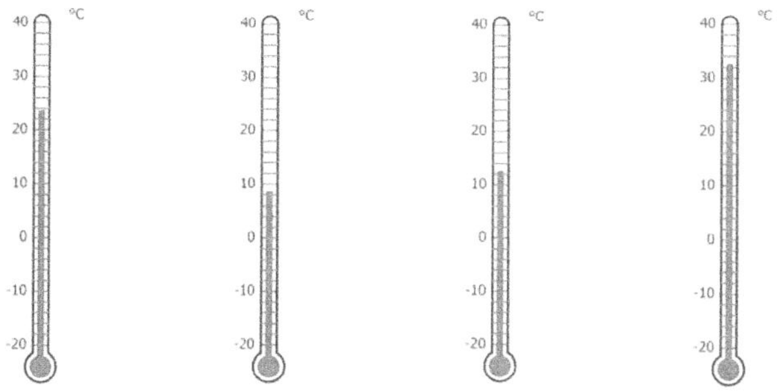

22°C   8°C   12°C   32°C

Q11. Page 150

a) 500ml

b)

Q12. Page 151

Yes, the line ends near 27cm.

Progress Check Page 151

Q1.

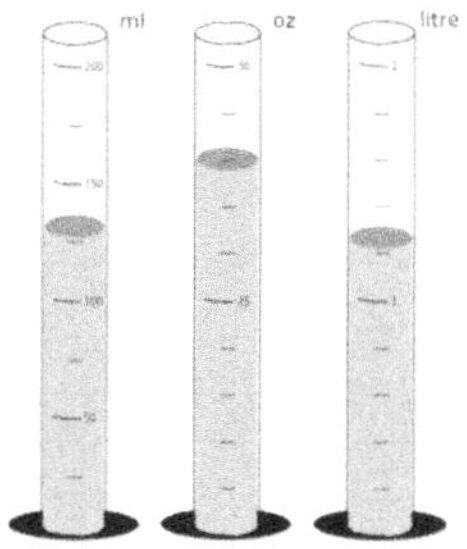

a)

125 ml   40 oz   1.2 litre

b)

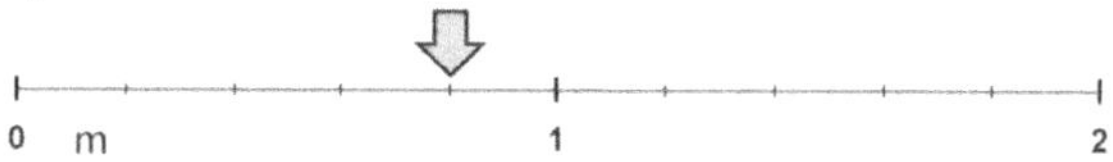

0.8 metres

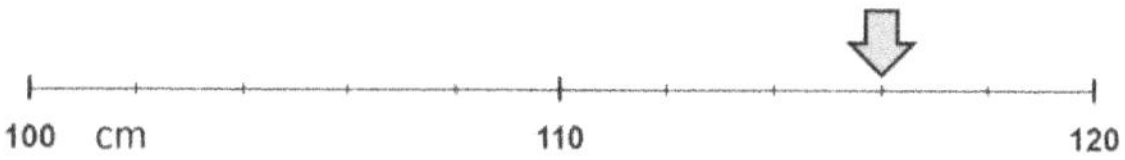

116 centimetres

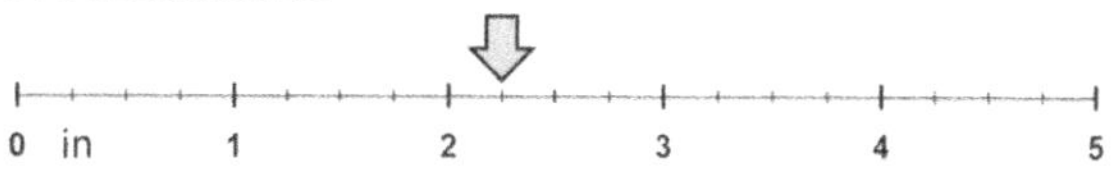

2.25 inches

c.i)

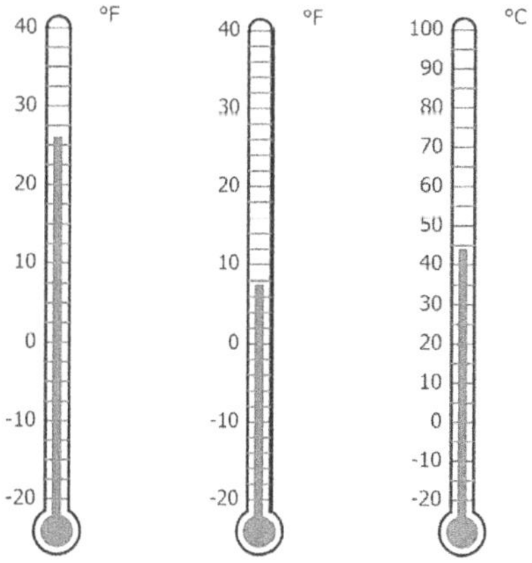

25°F   8°F   45°C

c.ii) 25 – 8 = 17°F

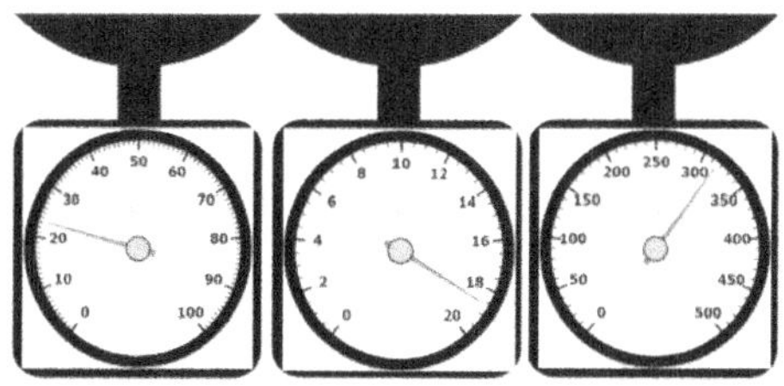

d)

23kg   18.5lbs   320g

e)
11cm

Sort 2d and 3D shapes using properties including lines of symmetry, length, right angles, angles including in rectangles and triangles

Q1. Page 152

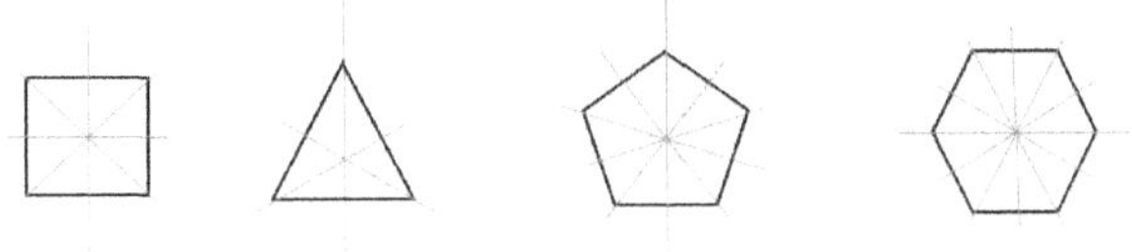

Q2. Page 152

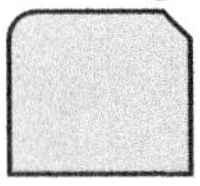

Q3. Page 153
a) square
b) regular hexagon

Q4. Page 153

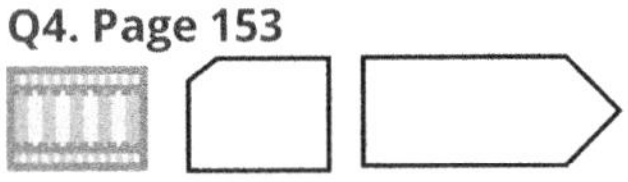

Q5. Page 153
rectangle – 4
triangle – 0
square - 0
hexagon - 0

Q6. Page 153
Squares and rectangles and some other irregular shapes.

Q7. Page 153

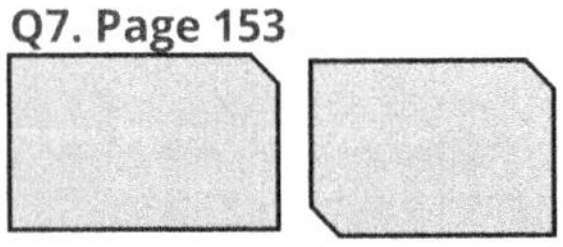

Q8. Page 153

Q9. Page 154
a) rectangles and squares
b) triangles
c) circles

Q10. Page 155
a) Cuboid and cube
b) Sphere
c) Cuboid and cube
d) Cylinder
e) 6
f) 3 (2 circular faces and 1 curved surface).

Q11. Page 155
Square, since the roof is a square-based pyramid.

Q12. Page 155
a) Cuboid
b) Yes, as it has right-angled corners.

Q13. Page 155
6

Q14. Page 155

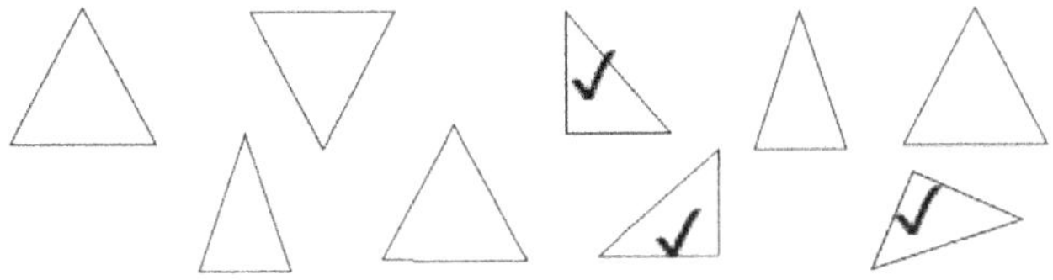

Q15. Page 156

equilateral triangle	scalene triangle	right-angled triangle	isosceles triangle

Q16. Page 156

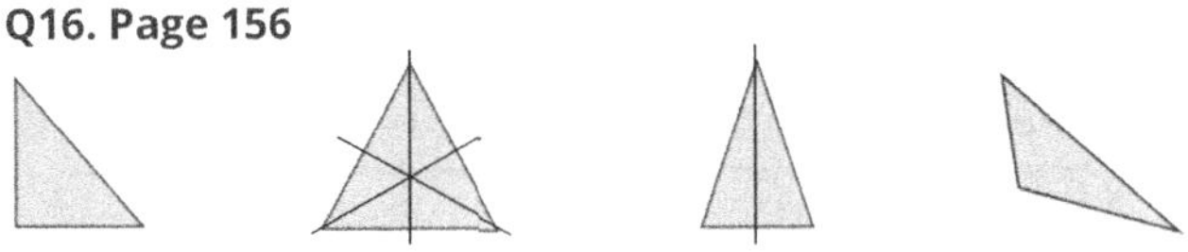

Q17. Page 156

Q18. Page 156

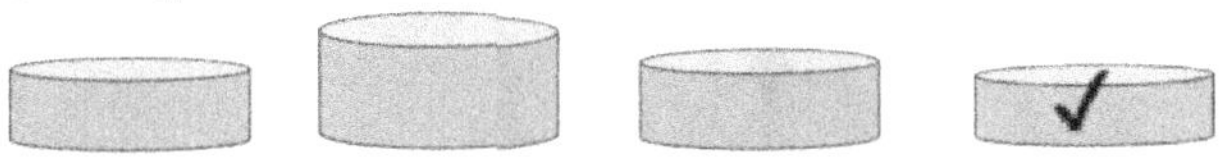

Q18. Page 156

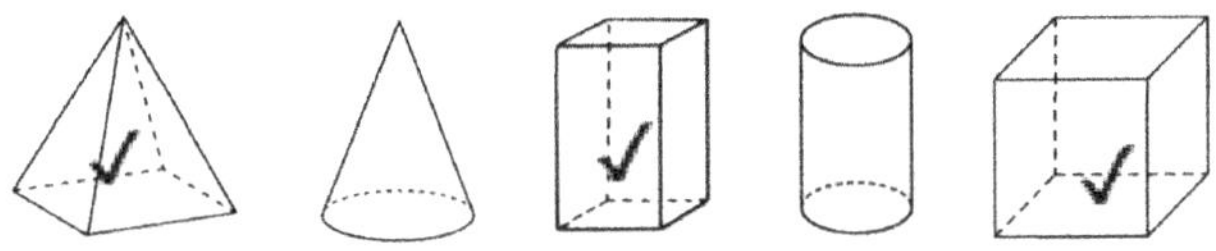

Q19. Page 156
a) Isosceles triangle
b) Isosceles triangle

Progress Check Page 157
Q1.

Q2.
a) **Hexagon** b) **Pentagon**

Q3.

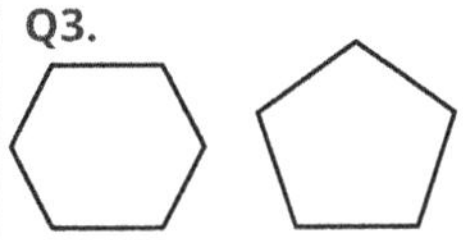

Q4.

a) He is not correct; a circle has an infinite number of lines of symmetry.

b) Not every triangle has three lines of symmetry. Only an equilateral triangle has three lines of symmetry.

William's thought about a sphere:

c) A sphere does not have a base at all; it is a three-dimensional shape where every point on the surface is the same distance from the centre.

Q5.

Q6.

a shape with a circular base.	a shape with 4 right angles.	a shape with sides of equal length.
		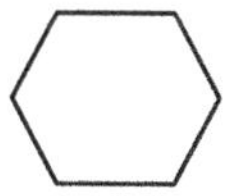

Use appropriate positional vocabulary to describe position and direction including eight compass points and including full/half/quarter turns

Q1. Page 158

North-west is in between **north** and **west**.
North-east is in between **north** and **east**.
South-west is in between south and west.
South-east is in between south and east.

Q2. Page 159

Everly Park is **north-east** of Lake Larson.
Ashton St is **south-west** of Bedford.
Snake River Park is **west** of Lake Larson.
Ocean Line Ave is **north-west** of Everly Road.
Green Park Ridge is **north-east** of Carlton Road.

Q3. Page 159

Nina is facing North. After a quarter turn anticlockwise, she is facing **west**.

Q4. Page 159

North-west from the supermarket.

Q5. Page 159

Dora is facing South. After a quarter turn clockwise, she is facing **west**.

Q6. Page 160

If Lynn has inserted some text upside down in a report, she needs to rotate it a **half turn** to correct it.

Progress Check Page 160

Q1.

SW stands for **south-west.**
SE stands for **south-east.**
NW stands for north-west.
NE stands for north-east.

Q2.

Bristol is in the west of London.
Cambridge is in the **south-west** of Norwich.
Exeter is in the north-east of **Plymouth**.

Q3.

S is **a half turn** from N.
SW is **a quarter turn clockwise** from SE.
NE is a **full-turn** from NE.
SW is **a half turn** from NW.

Q4.

Exeter, Devon, Plymouth or Bristol.

Q5.

Aaron thinks that the north is a quarter turn clockwise from the east. He is **not correct.** North is a half turn clockwise from the East.

Extract information from lists

Q1. Page 161

a) £23.99
b) garden tools set
c) the garden dining set
d) charcoal 12kg
e) £868.87

Q2. Page 161

a.i) June
a.ii) £120
a.iii) May
a.iv) 120 + 135 + 130 + 210 + 130 + 115 = 840

b.i) Year 11
b.ii) 131 + 144 = 275
b.iii) 127 + 131 + 144 + 150 + 165 = 717

Progress Check Page 162

Q1.

a) 11.99 + 0.50 + 0.50 + 0.50 = £13.49
b) Vegetable pizza
c) 14.99 – 11.99 = £3.00
d.i) 1.50 + 1.50 + 1.50 = £4.50
42 – 4.50 = 37.50   37.50 ÷ 3 = £12.50
She can only afford the vegetable pizzas.
d.ii) 3 x 11.99 = £35.97   35.97 + 4.50 = £40.47
42 – 40.47 = £1.53

Q2.

a) Weeks 3, 5, 7, 9
b) Week 9
c) 140 products
d) Weeks 1, 2, 4, 10
e) Week 3
f) 155 products

Extract information from tables

Q1. Page 163

a) Monday.

b) Tuesday.
c) Chemistry.
d) IT.
e) 263 - 76 = 187 more enrolments
f) 134 + 96 + 263 + 76 + 42 + 58 = 669
g) 58

Q2. Page 163
a) The destination that is **168 miles** away from the bus station is **City C**.
b) The destination that is **338 miles** away from the bus station is **City D**.
c) City D is **338 miles** away while City A is **155 miles** away from the bus station. The difference in distance is **338 - 155 = 183 miles** further for City D than City A.

Q3. Page 163
a) The temperature on **Sunday** is **20°C**.
b) The temperature is **16°C** on **Wednesday**.
c) The temperature is the highest on **Thursday**, at **22°C**.

Progress Check Page 164
Q1,
a) In winter 2021-22, **ten** 31 to 40-year-olds were referred.

b) The groups with the most referrals are **Under 15** and **15-20**, both with **60 cases** each.

c) There are **10 more** 21-30 year-olds referred than 31-40 year-olds (20 cases for 21-30 year-olds minus 10 cases for 31-40 year-olds).

d) To estimate the total number of referrals, add all the cases together: 60 (Under 15) + 60 (15-20) + 20 (21-30) + 10 (31-40) + 10 (41 and over) = 160 cases.

Q2.
211 + 357 + 452 + 335 = 1355 books.

Q3.
No, 127 + 151 = £278.

Extract information from diagrams

Q1. Page 165
a) Sue and Ali
b) Ava

Q2. Page 165
a) 24
b) Two
c) The kitchen is in between **the hall** and **the conference room**.

Progress Check Page 166
Q1.
a) Route 01
b) Two
c) Fireheart Canyon, Snowy Mountain, Emberfall Heights

Q2.
a) $\frac{1}{4}$
b) veggies

Extract information from bar charts

Q1. Page 167
a) 19
b) 23
c) Class A

Q2. Page 167
a) 92
b) 56
c) week 4

Progress Check Page 168
Q1.
a) Week 3
b) 82 – 59 = 23
c) Two weeks: week 2 and week 4

Q2.
a) Saturday
b) 320 + 290 = 610
c) 300 – 190 = 110

Extract information from tally charts and frequency tables

Q1. Page 169
a) 5
b) 4

Q2. Page 169
a) 10
b) 11 - 8 = 3
c) bus
d)

Subject	**Tally**	**Frequency**
walk	||||	4
bus	~~||||~~ ~~||||~~ |	11
train	~~||||~~ ~~||||~~	10
bike	~~||||~~ |||	8

Q3. Page 169

Service	**Tally**	**Frequency**
full groom	~~||||~~ ||	7
nail clip	|||	3
bath	~~||||~~ |	6

Q4. Page 169

Rating	Tally	Frequency
1	𝍷𝍷𝍷	3
2	𝍸 𝍷	6
3	𝍷𝍷𝍷	3
4	𝍸	5
5	𝍷𝍷𝍷	3

Progress Check Page 170

Q1.

a) 7

b) Cupcakes

c) 10 – 7 = 3

Q2.

Day of week	Tally	Frequency
Monday	𝍸 𝍷	6
Tuesday	𝍸	5
Wednesday	𝍸 𝍷𝍷	7
Thursday	𝍸 𝍷𝍷	7
Friday	𝍸 𝍷	6
Saturday	𝍸 𝍷𝍷𝍷	8
Sunday	𝍸 𝍷	6

Q3.

Type of book	Frequency
novels	25
non-fiction	15
nomic books	20

Interpret information, to make comparisons and record changes, from different formats including bar charts and simple line graphs

Q1. Page 171

a) 78

b) Yes

c) 78 – 65 = 13

Q2. Page 171

a) 120 – 110 = 10

b) January, February and March

c) January

d) The sales are going up.

Q3. Page 172

a) 2020

b) 63 – 52 = 11

c) 2021 and 2022

d) The number of books Gary has been reading from 2020 to 2022 has been increasing.

Q4. Page 172

a) metal

b) 49 – 38 = 11

c) glass in 2021

Progress Check Page 173

Q1.

a) 2019

b) Sales have increased in 2019 and then decreased dramatically in 2020.

c) 200 – 180 = 20

Q2.

a) Week 4

b) Increased in the second week, dropped in week three and then increased again in week 4.

c) 23 – 19 = 4

Organise and represent information in appropriate ways including tables, diagrams, simple line graphs and bar charts

Q1. Page 174

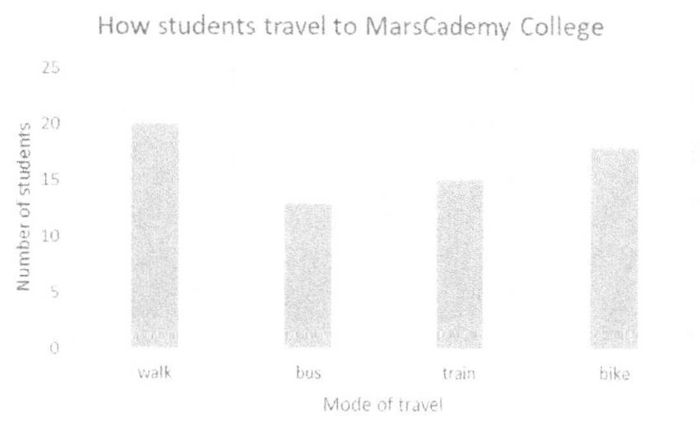

Q2. Page 175

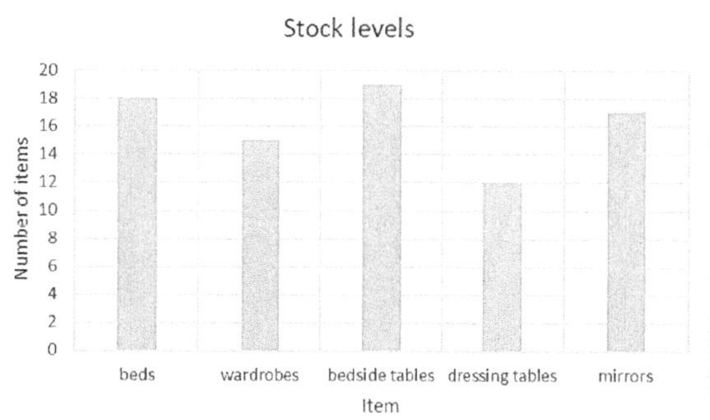

Q3. Page 175

Month	Number of items sold
Jan	78
Feb	81
Mar	85
Apr	81
May	90
Jun	93

Q4. Page 175

1. There is no title.
2. The axes are not labelled.
3. The scale is wrong

Q5. Page 176

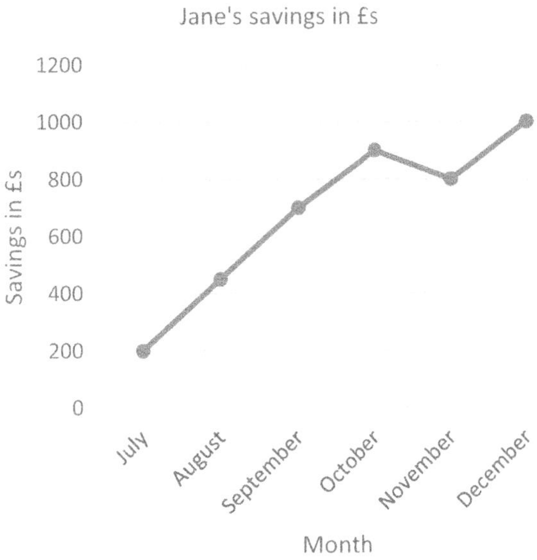

Q6. Page 177

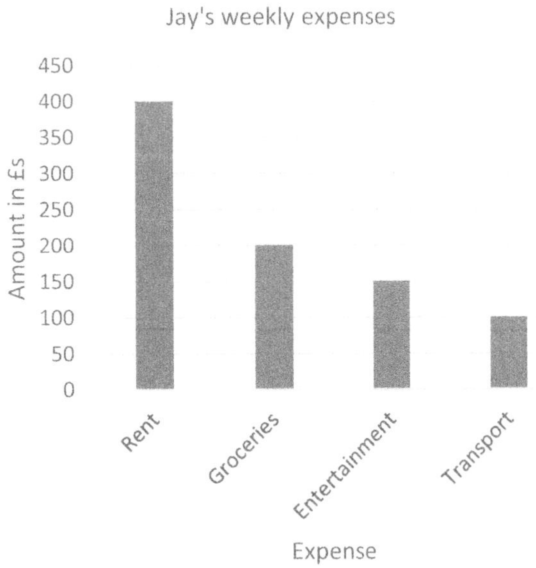

Progress Check Pages 178 and 179

Q1.

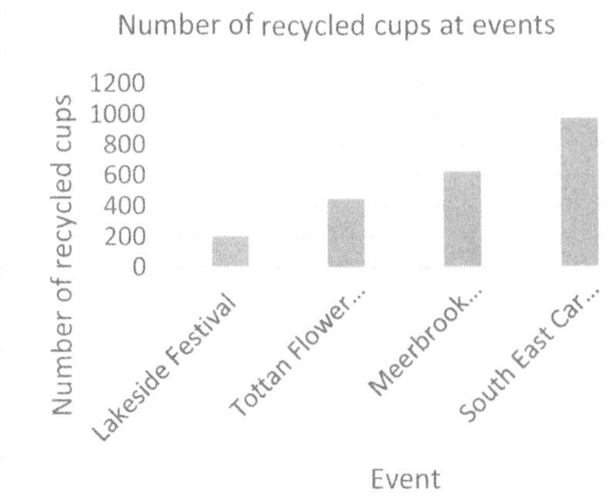

Q2.

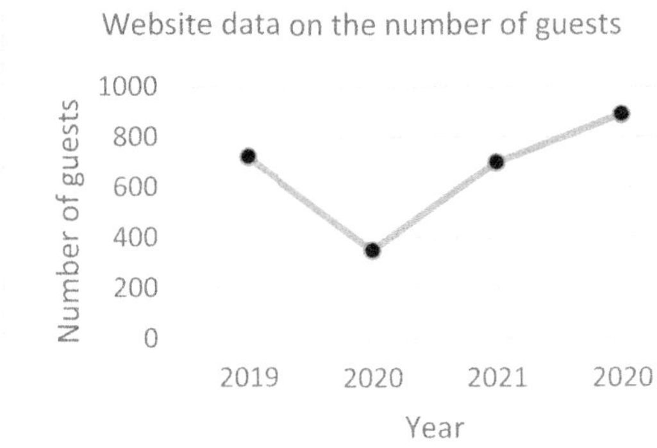

Practice Paper Part A Page 180

Q1. Three hundred and sixty

Q2.

215 + 75	(1 mark)
290	(1 mark)

Q3.

£50 – £45.49	(1 mark)
£4.51	(1 mark)

Q4. 245cm

Q5. Yes, he is correct. 15:30 in 24-hour time is 3:30 pm in 12-hour time. 15 – 12 = 3

Q6.

789, 625, 548, 301

Q7.

Q8.

Children's books.

Practice Paper Part B Pages 181, 182, 183

Q1. 750

Q2. 215, 295, 320, 400

Q3.

Q4. 3.6kg

Q5.

a) 480 ÷ 3

£160.

b) £480 - £160 = £320.

c) £320 ÷ 8 = £40

Q6.

a) 88 ÷ 4 = 22 pages.

b) Check: 88 ÷ 4 = 22.

Q7. The sequence increases by 4 each time, so the next number is 19.

Q8.

a) Line graph (example given) or bar chart.

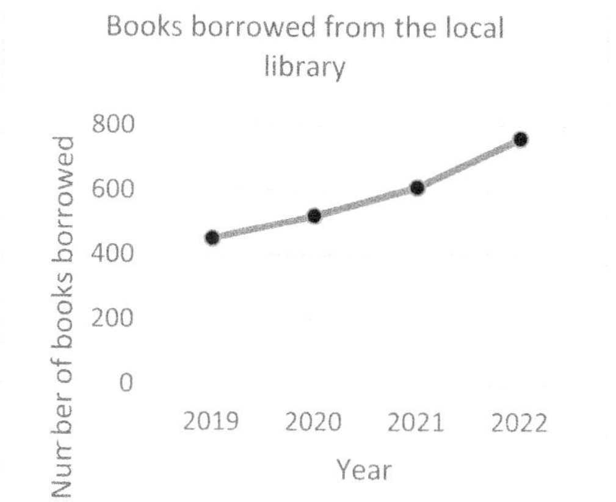

(adding the title - 1 mark, labelling the axes - 1 mark, plotting the points correctly - 1 mark)

b) 760 - 610 = 150

c) 2022

Q9.

14:45 - 13:35 = 1 hour and 10 minutes.

Q10.

a) £1.50 + £0.75 + £0.75 + £1.50 + £2.25 = £5.25. (1 mark)

Change from £10: £10 - £5.25 = £4.75. (1 mark)

b) 4 customers

Q11.

Colour	Frequency
red	6
blue	5
green	4
yellow	4

(writing all the colours in the Colour column - 1 mark, writing two of the frequencies correctly - 1 mark or writing all four frequencies correctly - 2 marks)

Q12.

945 ÷ 13 = 72 remainder 9

Q13.

No, this is closer to 50ml.

Q14.

South

Q15.

a) 76 ÷ 4 = £19

b) 80 ÷ 4 = £20

Q16.

£18.60

Getting 24 marks or more in the practice paper, means you have passed it.

If you don't get 24 marks or more, go back and revisit topics you lost marks in and retry the paper.

More practice papers can be found on **marscademy.com**

Made in the USA
Monee, IL
13 September 2026

62975578R10118